Blackout: A Look Inside Wernickes

David J. Steele

© Copyright 2011, David J. Steele. All rights reserved.
ISBN: 978-1-105-4395-6

This book is dedicated with gratitude to my friend Karen East, who works with people suffering from Wernicke-Korsakoffs.

David J. Steele

There's a man in a hospital bed in neural intensive care. He's thin, dark-haired, and in his late thirties. A short man, he weighs a hundred and thirty-nine pounds. Two IV's are stuck in his left arm. His left wrist is strapped to the rail in a soft restraint. He appears to be asleep. It's deeper than sleep, and has been for almost twenty-four hours.

That's me. That's where I was in July 2005.

He opens his eyes. Blinks. The environment isn't new to him, but he doesn't know that. As far as he's concerned, his name is Tom Benton and he's a prisoner of the Protectors Guild of Sexton. He's a character in a novel he started twenty years before he woke in that bed, and that's not good.

With a twist of his shoulders and fumbling hands, he frees himself. He sees needles in his arms. They bother him. He pulls them out one at a time. Looks around the room as if he expects to be attacked. When no one comes, he sits up in the bed and climbs out. His gate is unsteady as he walks out the door.

Don't worry. He won't get far before he blacks out again. Next time, they'll use a different restraint.

§

That was six years ago. I'm fine now. Sober. I enjoy being sober more than I ever enjoyed drinking, and that's saying a lot. I was a beer guy. I drank other alcoholic beverages, but my first choice was beer. I tried to turn it into a food group, a staple of my diet. Using drunk's logic, it seemed to me beer was liquid bread. (It's okay to smile from time to time as you read this. Humor gets people like you and me through a lot.) I used to make beer, and it was easy for Drunk Dave to imagine it had nutritional value: grains, yeast, water...stuff in bread. Why bother with taking wort pills when you can just drink a beer or two, or twelve? Drunk's Logic.

I became adept at drunk's logic, to my detriment. It's easy to look back and see the drunk's logic now, but believe me, it seems sound when you're caught in it. I offer no excuses. I'm also not going to offer a lot of detail about how and when I drank. Part of the reason I'm not going to offer a lot of detail is shame. I feel badly about my drinking. I don't tell you that in an attempt to find absolution from the guilt. I need to feel guilty about my drinking. It's part of how I stay sober. The other part of the reason I'm not going to share how much I drank is that I don't want people to hold my level of drinking up as any sort of measure of their own and think they're safe. I did that when I visited Alcoholics Anonymous meetings: compared how much I drank to how much whoever was talking said they drank, and from there figured I was okay. It doesn't work that way.

I drank a lot of beer. Every day I drank something, usually beer, but sometimes mixed drinks, sometimes wine, sometimes all three. People tend to have an image in their mind of alcoholics being continually drunk, but that's not always the case. I didn't drink to get drunk. I drank to get a buzz. I drank to relax. I drank for the taste...drank to get sleepy.

Wernickes is a nutritional deficit. Thiamin—Vitamin B1—doesn't dissolve in alcohol. With a belly full of beer, that stuff is going to sit around for a while, get bored, and go away.

David J. Steele

Other bad things happen to brain and body with excessive, chronic alcohol abuse. I'm not sure what they all are, but doctors and others can tell you. For the purposes of this story, you really only need to know a couple of things: I was a heavy drinker, I succumbed to Wernicke-encephalopathy, I have Korsakoffs, and I'm sober now.

You might ask yourself, as I'm asking myself while I sit here at eleven PM on a Saturday night sipping a mug of decaf... If you're not going to tell us about your drinking, Steele, what *are* you going to tell us?

I'm going to tell you a story, nothing more and nothing less. Ready? Of course you are. I'm not sure *I* am, but here we go...

Section One – Prelude to the Storm
I
Spring day, 1993

I was a hot-to-trot executive with the Boy Scouts. I was good at what I did. I was a good fund-raiser, I could recruit and lead volunteers, and I was passionate about the program. I was ambitious in a good way. We had a new director, a man I like a great deal and who demonstrated a lot of faith in me. That meant I was working a lot. Eighty-hour work weeks were my normal. When I wasn't working, I drank beer. I was twenty-seven years old and was only a few months away from another promotion. I was recently married. Still married to the same women, and as you'll see later, I'm lucky to be married to the same woman.

There was a pavilion outside the office. The day was bright and sunny, and it was late enough in the Michigan spring that the first dandelions of the season were blooming a yellow rash in spots around the lawn, which was thick and green and ready for the first cutting of the season.

I walked out of the office in the late morning sun and took a seat on top of one of the picnic tables under the pavilion. With my polished black shoes on the bench and my blue suit-clad tush on top of the table, I put my face in my hands and rubbed my eyes. I was seeing a yellow blur flicker in the corners of my vision. I thought it was there because I was tired. I thought it was an ocular migraine.

It wasn't my first one. When I got tired, which was most of the time, I felt the skin under my eyes twitch. Sometimes, as then, I saw glittering shapes at the outer edges of my sight. There was no pain. Glittery, spangle shapes, floating in my glance. If I gave them time, they went away. I smoked a cigarette and waited for my vision to clear.

I remember thinking at the time that it might have something to do with the beer I drank. It was easy to shrug away the thought. Too easy. By the time I finished the cigarette, the light was gone from my vision. I went inside, tossing the butt in the can by the door, and went on with my day.

II
May 29, 1993

It was our first wedding anniversary. We were engaged for two years before we married and had known each other for a little over three years by then. She graduated from college just before we got married, and although she hadn't landed her first full-time teaching job yet, she was a substitute teacher who worked almost every day, and she had a part-time job as a clerk in a shop in town.

The plan was that we would go out for a nice dinner when she got home from her part-time job. It was a Saturday, a rare

David J. Steele

Saturday for me because I didn't have any appointments that day. I got up in the morning and worked on a quilt I was making. Yes, I'm a man and I like to make quilts. The one I was working on was a serious undertaking. It was a pattern I lifted from a counted cross-stitch pattern. A knight riding a pegasus. Instead of little stitches in a hunk of cloth, I cut out squares of fabric and sewed them together by hand.

I drank beer while I did that. Drank beer, stitched one-inch squares of fabric together, and watched television. I had worked three weeks straight without a day off before that day, and I was tired. About two o'clock that afternoon, I decided I needed to sleep for a few hours. That was my sad *modus operandi* for quite a while: get up, do stuff, drink coffee until about noon, switch to beer, sleep for a while, do some stuff, drink beer, go to bed. It's not a healthy cycle, by any stroke of the imagination. I make no excuse for it.

It was getting dark when I woke up. My wife was in the bed next to me, sleeping. I woke her up and said it was time to go out for dinner. She wasn't happy with me. Pretty mad, in fact. She said she tried to wake me up several times, and when she finally got something out of me, all I did was ask her to join me for a nap. She had been home for six hours, which meant I was asleep for eight hours.

I chalked it up to being tired from working for three weeks without a day off. I know now that I was wrong. I was in a deeper sleep than normal.

We went out for dinner. I didn't eat much. I wasn't sick, and I didn't feel drunk although I'm sure I was. I felt funny: kind of out of place. My sight was dim. I thought I might be coming down with something. I think it might have been a small bout of Wernickes, if such a thing is possible. I read somewhere that those who come through the most serious part of the disease might have had it before the 'big one', but the researcher didn't have documentation. Wernickes seems to be as under-reported as it is under-diagnosed. Those are my non-medical opinions based on a lot of reading.

III
August, 1997

I was promoted, and the promotion involved moving to one of the far suburbs of Chicago. It was a big change, and my wife and I were excited. The move went smoothly. We found a four-bedroom house in a subdivision about twenty miles from my office. Sarah found a job right away, teaching in a school district West of us—which meant she didn't have to brave Chicago traffic on her way to and from school. I had an administrative assistant, and a staff of five professionals to supervise. The house in Michigan sold for what we paid for it and I my former boss back there closed on it for us with power of attorney.

It seemed like life was good. Looking back on it now, I can see I wasn't very happy. I didn't enjoy the job as much as I thought I would, but I didn't admit that to myself. I was swathed in a couple of layers of denial: I didn't want to admit my drinking was getting out of control, and I didn't want to admit I was losing my love of my chosen profession. The timing of the move was good. A couple of days after I started the new job, the first national convention of all BSA professionals started. It was a good chance to get to know my old team while spending some time with my old friends.

...Those things were the farthest things from my mind when we boarded the plane at O'Hare to head to Nashville.

It was a convention of Boy Scout professionals, held at the Opryland Hotel. If you've never been there, you go see it. It's

David J. Steele

a glorious hotel. It's huge. It's enclosed in a dome. There are plants and trees, and a river, and a boat, and stores, restaurants, bars, and more. We were going to be there for five days and had no plans to leave the hotel. It's that big—you can have an entire vacation without knowing you never set foot outside.

I was impressed with the place, and we were having a great time. It had only been three weeks since we moved. I enjoyed my team, the team of professional Scouters I'd joined, and I missed the team I had just left. I was sharing a room with my former supervisor, but he was more than a supervisor. He was, and is, an excellent friend of mine.

He brought me a copy of a newspaper story that ran when I left Midland. Midland isn't a big city—it has a population of under 40,000. Great town. I loved the time we spent there. It's a Scouting town. One out of three boys of Boy Scout and Cub Scout age were members of the Boy Scouts of America, and I had the privilege of being their executive for seven great years. I read the article several times and looked at the picture—amazed that a newspaper would print a 1/4 page article about me.

The council I was serving had a suite for the night. It was on one of the highest floors of the hotel. Huge windows overlooked the miniature world that is the Opryland Hotel. I drank a lot that night. We all did. People came and went throughout the night and those of us hosting the party made sure they all had a good time. The party was showing signs of winding down at about one AM and the boss's wife let me know she wanted me to clear the room out.

I cleared my throat and said, "It's late and we all have a meeting tomorrow. I expect to see everyone at breakfast!" There were some good-natured boos and some ribbing back and forth. I remember that clearly. I meant what I said…

Blackout: A Look Inside Wernickes

The next thing I knew, I opened my eyes in the hotel room. Someone was shaking me. It was Steve. He pulled back and shouted, "It's two o'clock. I thought you were up."

"Up?" I shook my head to clear it, but it didn't work very well. I looked around.

The balcony doors were open on the interior of the dome. I could see bright light, sunlight from the skylights, and green plants out the balcony. I rolled over and looked at the alarm clock on the nightstand. It was two o'clock in the afternoon!

So much for beating them all to breakfast. So much for the opening session. So much for...

"You must've gone back to bed."

"I wasn't up."

He grinned at me and tossed some stuff on his bed. I glanced at the stuff. It was a pile of the usual conference stuff: brochures, give-away pens, notebooks, etc. Stuff I should have been gathering between sessions. "You didn't miss much. Your staff was looking for you." He started to head back out. "Oh yeah... You won a trip to Branson, air fare included. Weekend stay, or something like that. One of your team members claimed the prize."

"Where the hell is Branson?"

"Missouri. Second country music capitol of the world."

"Damn," I muttered. "Punishing me already, aren't they?" I'd rather have an unexplained rash than listen to country music. The joke was wasted. He was already gone.

I didn't feel well. My head felt funny, and not just with a hangover. I felt like I could have rolled over and gone to sleep for a couple of days. I jumped out of bed and started to pull clothes on. Guilt and worry hit me when I was in the shower. I was sure I was in trouble, and thought I deserved it. My new boss wouldn't be happy about me missing the first couple of sessions, and getting drunk and sleeping most of the day was far from the kind of example I wanted to set.

I had to live with the worry for a couple of hours. By the time I ran into my boss, I felt pretty normal. Guilty, but normal. He was on his way up a set of marble stairs leading to one of the big ballrooms used for a general session. He grinned when he saw me.

Before he could start yelling, I stopped him. "I overslept," I said, "…by a lot. I'm sorry."

He grinned, and waved to a couple of people standing at the top of the stairs, looking over the waterfall and river on the other side of the stairwell. "You've been through a lot in the last few days: moving, unpacking, getting used to the office." He slapped me on the back as he went by. "See you at dinner."

IV
January, 2001

In most fiction, there is a scene in which the protagonist could avoid the trouble that makes the story. This isn't fiction, but it is that moment…

I had just come back from lunch and was hanging my coat up on the back of my office door when the boss's administrative assistant stuck her head around the door.

"He'd like to see you when you get a chance."

She vanished before I said anything. I didn't think much of it at the time. I was in a pretty good mood. There was no reason I shouldn't have been in a good mood. We were at the start of a new year after a successful year of membership gains, goal accomplishment, and fund raising. I felt good, too good, after a couple of beers and as many bites of a hamburger for lunch.

I walked around the corner to his office and was surprised

to see him sitting at the table by the windows. He was a big guy: broad shoulders, round face—the kind of face that could make a grin contagious—and expressive eyes. He wasn't smiling. There was a manila folder in front of him on the table. It was a thin folder without much in it.

"Take a seat," he said with neither smile nor preamble. "Close the door first."

I don't remember the exact details of the conversation, but I remember it felt surreal.

"Have you ever heard of the E.A.P?" he asked.

I said I hadn't, but he clarified for me. It was in the employee handbook—the one I just helped revise. I had no excuse for not remembering it.

"It stands for 'Employee Assistance Program'. You have a drinking problem. I hear things…"

I didn't say a word. My mouth was dry, and my heart was beating too fast. Excuses piled up behind my teeth, but I didn't bother to utter them.

He pulled a card from the file and slid it over the table to me. "This is the number for the referral program. It's an eight hundred number, and they answer it twenty-four hours a day. You're going to call that number, and you're going to do what they tell you to do. If you don't call that number in twenty-four hours, I *have* to fire you." He leaned forward, locking his eyes on mine. "I don't want to fire you. You're too valuable to this organization. If you weren't valuable, I would have already fired you. Call that number, Dave."

"I'll call."

I called that night. Hands shaking, ashamedly buzzed from a couple of beers to help me gather the nerve (a moronic move, an *addict's* move), I dialed the phone hoping it wouldn't be answered. It was answered. A very understanding person informed me they were expecting my call. She told me it was an employer referral. I asked what that meant. It meant, she explained, that my employer was sending me through the Employee Assistance Program and that in order to participate, I would be required to sign a waiver that would allow them to share information with my employer about my participation and progress in the program.

I stared at the phone for a long time after I hung it up. I drank a few more beers before my wife got home, knowing I would have to tell her about the phone call I just made. By the time she got home, I managed to convince myself it was all some sort of mistake. I wasn't an alcoholic. I was just a guy who liked beer. Liked it a lot, to be sure…but I just liked it. She would laugh and tell me to go through the program.

She didn't laugh. She thought it was a good idea. It was, of course. I know it now, you knew it when you started reading this, but I didn't *want* to know it then.

I went through the program. Nine weeks of alcohol education, mostly attended by people who had DUI's on their records and were required to attend the class before they could start the process to get their driver's licenses back. I stayed mostly sober through the nine weeks. When it was over, I went back to drinking. I drank at a slower pace, but I kept drinking.

V
Thoughts before we get to the important stuff

I started drinking beer again after I got through the program. That doesn't mean the program wasn't good, and that I didn't get any help from it. I didn't use what I learned, not until after I got home from the hospital five years later. During the nine weeks, I learned I could handle sobriety. I could have learned to *enjoy* sobriety, but I didn't. I also learned

I didn't have to rely on beer to get to sleep. I took melatonin then. You'll hear more about that later.

In 2003, my wife and I moved to Racine, Wisconsin. It wasn't a promotion. I took a job as the number two professional in the Boy Scout council there. We liked living in Wisconsin. A year and a half later, we had a budget shortfall that was only a little bigger than my annual salary. When you're the number two guy in any company and the budget shortfall matches your salary, you know what you do? You pack your bags.

It might surprise you to know I had a very good performance record, in spite of my drinking. Several other middle managers in the organization were laid off, or otherwise displaced for budgetary reasons that year, and a lot of good professionals left the organization. I thought about leaving. I was vested in the retirement program, but I loved, and still love, the Boy Scouts of America and I wanted to hang on. The organization worked on my behalf. The day I was told my position had been eliminated, I called the Regional Personnel Director. We knew each other. Not well, but well enough that when I said...

"I didn't think I'd be calling you like this on my birthday..."

He said, "Wait a minute! It's your *birthday?*" He laughed. Hard. Very hard. It was such a surprised, happy sound that I laughed with him. Laughed with tears in my eyes, but I laughed. "They laid you off on your freakin' *birthday?*"

"Well," I said as I coughed through laughter and burning eyes, "I didn't tell them it was my birthday."

That brought another spate of laughter. I couldn't help it. I laughed too, and felt better for it. I was sitting in my car in the parking lot on a beach on Lake Michigan on December 20th.

Wind was whipping over the car. I watched light snow blow around the parking lot.

"We'll find something for you, Dave." He wasn't laughing when he said that, thank God. "You have a great record, and we'll find something for you. I have to warn you, it might not be a promotion. There are a lot of guys in your position. A lot of councils hit a financial hard place this year, and lots of positions have been eliminated. Don't quit! We'll find something for you."

They did. I made phone calls. My boss made phone calls. The region made phone calls.

I got a job as a very senior district executive. It was in Cleveland, Ohio. The executive director there offered me the same salary and job classification I had in Wisconsin. It was a demotion in title, but the council there was looking to create management positions, and if all went well, one would be mine.

My wife took the news very well, in spite of never wanting to live in a big city. We always agreed that she wouldn't move during a school year and until then we were able to time our moves so we moved in the summer, but not that time. That time she stayed in Wisconsin, and I got a furnished one-bedroom apartment on a six month lease in Cleveland.

Was I smart enough to be sober when I moved to Cleveland and waited for her? We know the answer, but I'll say it anyway.

I wasn't.

Section Two – The Storm
§
Intro to *Green Goblin*

Looking back on it, I see Wernickes coming on. I didn't see the little warning signs then. Some medical professionals will say there aren't any warning signs. Others will say the signs are often missed, and those are the ones I agree with. I've read some stuff that says Wernickes comes on suddenly. I don't know which ones are right, and which ones are wrong. I think some of the things I've described so far are indicative of Wernickes.

I was also having problems with my memory, but I didn't know or believe I was having problems with my memory. My wife, bless her, told me I was forgetting things, but I was able to cloak her assertions in denial.

Then I got sick. Really sick. I had breakfast the other day with an old friend from camp staff. He said he was surprised at my recovery. My father told him I was ill when I was in the hospital for Wernickes. This is what my friend said:

"Your Dad told us the hospital said you probably weren't going to make it. When I saw him a couple of weeks later, I was almost afraid to ask. Your Dad said you were home and doing fine."

I was lucky. I was blessed. I didn't know how lucky, or how blessed, until much later.

The next part of this book is a book on its own entitled *Green Goblin*. It's available as a Nook book, a Kindle book, and in hardcover and paperback from www.lulu.com/spotlight/Misticuf. You don't need to buy it, but if you want to send it to someone else, it's a good cautionary tale.

§
Green Goblin

Werknicke-Korsakoff Syndrome: *A disorder of the central nervous system characterized by abnormal eye movements, incoordination, confusion, and impaired memory and learning*

David J. Steele

functions.

Wernicke's Aphasia: *A type of aphasia caused by a lesion in the Wernicke's Area of the brain and characterized by grammatical but more or less meaningless speech and an apparent inability to comprehend speech.*

Part I
Bed Sheets and Brimstone
1

The journey I was about to take was both long in coming and a surprise. Recently I read description after description of the illness I suffered. I'll tell you what it is at the end of this story. If I remember. If there is an end to this story.

It began with a dream.

The Klingons were blowing up the *Enterprise* and there wasn't a damn thing I could do about it. No one responded to my orders to fire the phasers and I have no idea how to run a starship. The deck shook. I turned to Worf and ordered him, again, to fire. He wasn't there—no one was.

My pulse hammered in my neck. I was scared but determined to go out fighting. Then I saw her: my wife Sarah, dressed in Deanna Troi's uniform. She was crying and giving me a strange look. I wanted to hug her and tell her it would be okay. We'd beat the bastards but I needed her to fire the photon torpedoes and do it *now*.

"Don't stand there and cry! We're going to die if you don't fire the torpedoes!" I hated myself for shouting at her. I don't shout. It's not my style. Of course, getting killed by Klingons isn't my style either.

Then I woke up. I was lying on the floor on her side of the bed, pulling myself up with the quilt. "It's okay," I said. "I'm awake now."

She looked scared. Sarah doesn't looked scared any more often than I shout. She told me to get dressed. We were going to the hospital.

Blackout...

2

I stood on the driveway on her side of the garage. Sarah was behind the wheel and we were waiting for the light to turn green. I was in the passenger seat looking at First Baptist Church through the window. I was an executive with the Boy Scouts of America at the time and we had several meetings a month at that church.

"I don't think I can go to a meeting tonight," I said. "I'm not in uniform."

Sarah sounded calm. "We're going to the hospital."

She had been fighting colitis for more than two years and her symptoms led her to the hospital only a couple of weeks before.

"Why are you driving?" It didn't make sense for her to drive herself to the hospital.

"You're drunk."

She had me there. I was drunk. I drank too much anyway, but for the couple of weeks prior to the Klingon invasion, drinking large amounts of beer was the only way I could find to get rid of the double vision.

I saw double for at least two weeks before I went nuts. I should have gone to the doctor in early June, but it was a busy time of year for us. I felt fine physically so I didn't see double vision as a high priority symptom. Instead of going to the doctor like I knew I should, I got new glasses. I still saw double, but I saw double *clearly*. I kept an empty foam cup in my car for the random bits of puking I experienced. There was a lot of random puking, but it was okay. There wasn't much to throw up. It's hard to eat when you see double and puke every now and again.

But I didn't need to go to a *hospital*. I was fine. She was the sick one.

Blackout…

3

I was in a hospital trauma center waiting room very late at night. I didn't know how I got there. Sarah was filling out paperwork at the counter. I wanted a cigarette. I reached in the left pocket of my jeans, but for some reason I didn't have my cigarettes.

I started to walk to the car. It was the first of July and a nice night—warm but not hot. The parking lot was almost empty. They usually are at 2:00 in the morning. I turned around and went back thorough the emergency room entrance. I thought Sarah would be admitted soon and didn't want to bother her for her keys. My plan was to walk home, get my cigarettes and drive our other car back.

On my way back from the lobby I saw three or four wheelchairs and decided to take one home instead of walking. I heard Sarah shout and try to catch up with me, and thought, she doesn't have to worry about me. They'll help her soon.

"Wait!" she shouted.

I waved over my shoulder. "See you at home dear!"

So there I was, rolling my wheelchair toward Mayfield Road at two o'clock in the morning, going home to get my cigarettes and my car. I stopped before I turned left on Mayfield. Even at that hour, there were cars on the road. Safety first!

I'm a good driver and try to be polite. I intended to use my turn signal, but couldn't find it on the wheelchair. It seemed like a bad idea to get killed in the center lane while driving a wheelchair. It didn't take long to figure out wheelchairs don't have turn signals. I started to roll out in the road anyway when I heard someone running behind me.

I turned. Two men were gaining on me. One was bearded and big. He was in front of a short, stocky guy. They grabbed the wheelchair just as I threw caution to the winds and went for the road. I struggled, but one held the wheelchair while the other grabbed me.

I told myself not to hurt them. They weren't hurting me; they were restraining me. It sounds funny, I know. I'm a little guy—five foot six and a hundred twenty pounds—but size doesn't make any difference in some situations. What matters is who walks away and who doesn't.

I stopped struggling when I realized I couldn't get away without hurting them. Better to wait and see what happens.

They rolled me back through the doors. Sarah hugged me just before they took me into the patient storage area—you know, where they put patients until they get around to fixing them. She was crying, I was confused.

So ended July 1, or so I thought. It was actually the wee hours of July 4th—I'd been unconscious at home for at least thirty-six hours since leaving work for vacation on the first. *Blackout...*

4

I opened my eyes and found myself looking at a woman's face. She was a blond gal, neither ugly nor pretty. She was talking to me, but I didn't understand what she was trying to say. It's funny, but I was more curious about why I was sleeping sitting up in bed than why I was there.

"You have to sign this." She put a clipboard on my lap.

I looked at the piece of paper on the clipboard. It was broken in block paragraphs. There were a couple of lines at the bottom for signatures. I tried to read the document. I don't sign anything without knowing what it is I'm signing.

I tried to read it. I couldn't.

"Don't you have one of these in English?" I laughed but I wasn't kidding. Not one word made sense.

"It's in English." She looked worried.

I looked at it again. I'm pretty good with languages. I can read some French and Spanish. I recognize German, Russian, and Hebrew when I hear them. I didn't recognize the language on the paper in my lap.

She didn't know what was holding me back. She sounded desperate when she said, "You *have* to sign it! Your wife has been calling me every day! She's very worried about you. She loves you very, very much."

I looked at the form, then at the nurse. I wanted to sign the document…but I didn't know how.

"We can't tell her you're here if you don't sign it. It breaks my heart to hear her cry. Please sign it so we can tell her you're here."

I looked at the document again. It was hard to think, but I found the words I was looking for. "How do I sign it?"

She looked surprised, then more worried. For the first time I noticed a man standing behind her. I didn't see his face. I remembered Sarah took me to a hospital. If these people couldn't or wouldn't tell her where I was, that meant I'd been taken from the hospital. *That* meant I was a hostage.

Kidnapped? Me? If they wanted a big ransom for little old me, they could forget it. We didn't have money to pay a ransom. I decided to get out of there, but first I had to sign the form. Play along and watch for an opportunity to escape. "I don't know how to sign it."

She talked me through it. I formed the letters as she said them. By the time I wrote "David J. Steele" in childish cursive letters, I remembered how to write. I tried to hand the clipboard back to her, but she pushed it toward me. "Please date it."

I nodded and stared at the blank. "What's the date?" What she said didn't make sense to me, but I wrote it anyway.

"Seventh floor," she said.

"It's a four-story building."

She looked at me an enunciated every syllable. "Sev…en. … Four."

That made less sense than seventh floor. So I wrote seventh floor hoping she would accept it and let me go back to sleep. I had an escape to plot and needed to rest before I executed whatever I came up with.

I heard her say to the man behind her, "I think that's as good as we're going to get."

Blackout…

5

I felt a sharp pain in my arm and opened my eyes. There were two men holding my left arm. The shorter one, the one at my shoulder said, "He's awake!"

I yanked my arm free by pulling it toward my chest. They let go. I bent my elbow and thrust for the throat of the man closest to my wrist. His face changed as my hand got closer to his neck. He looked happy, then surprised. I pushed my hand closer. My first thought was to crush his windpipe. Push the hand; push it toward his neck. Squeeze—crush, don't choke—crumple his esophagus like a beer can.

That scared him. He tried to force my hand back, but I wouldn't let him. We struggled. He couldn't push my hand away and I wouldn't stop reaching for his neck.

I didn't want to kill him. I only wanted to injure him, severely if I had to, and hoped his partner would care more about saving a coworker than the little prisoner in the bed. I intended to pluck his adams apple from his neck and hold it up for the other guy to see.

It doesn't take long to recognize a bad plan. I couldn't get to the man's neck. I gave up on the idea of ripping his throat apart and switched to escape mode. I lashed out with my right arm to grab the rail on the other side of the bed. Grab the rail, pull hard, and vault out of bed. Then run for the door and get the hell out of Dodge.

My hand slammed on a woman's wrist resting on the rail. I heard a choked cry and looked up. She was wearing a white coat. I didn't look at her face; I looked at her hand.

Think! I needed time to think, but didn't have it. Stalemate. The men had me and I had the woman. I didn't want to take a woman hostage, and I didn't want to hurt her, but I hated the idea of getting killed more. Since when did the bastards in the Protectors Guild use women? There were no female guildsmen the last time I was in this world.

I had to do something. It went against the grain for me—hurting a woman. I didn't want to do it, but I knew she would slide a dagger between my ribs with none of the hesitation I felt toward hurting her. I wanted to tell the men I would let go of her if they let go of me. The words wouldn't come. I couldn't say anything…I didn't know how.

I squeezed her wrist. In my mind's eye I could see her bones just above my curled fingers, the twin, not quite parallel bones in her forearm. My fingers were clenched. *Squeeze. Tighter.* Pop the hand off her wrist like the flower from a dandelion! I locked eyes with the man holding *my* left wrist. I wanted him to look in my eyes and see I would rather kill him ten times than lie in that bed as their prisoner. As I did that, I squeezed the woman's arm harder. She would tell them to let me go any second. I smiled at the Guildsman to intimidate him.

"David! Stop it!" the woman shouted. "David! You're *hurting* me! Please...stop!"

...That was the last thing I expected her to say. *She sounds like my Mom!* No one but my mother calls me David. Guild bastards. How was I supposed to hurt someone who sounds like my mom?

I couldn't.

I let go of her arm and waited for them to kill me. The last thing I heard was a man's voice. "The little fucker smiled at me! Did you see that? He tried to kill me and he fucking *smiled!*"

Blackout...

6

I woke up when I heard someone walk in the room. I opened my eyes to a woman standing between the bed I was in and an empty bed next to it.

"You're awake!" she said. "How do you feel?"

I felt fine and said so. I didn't know where I was or why I was there, but the bed was comfortable and I just woke up.

"You look terrible," I said. "You should sit down. She sat on the other bed and gave me a warm smile. I was looking at a woman in her mid-fifties. Slender, attractive. She had dark hair with a few strands of gray. I liked it when she smiled; it was a nice smile. She looked me over from her perch on the other bed. Her expression held curiosity and surprise. She seemed very relieved about something, but I couldn't tell what. "You *are* a nice man. You're a *very* nice man." She smiled again. "I told them you were, but they said you were dangerous. They said you tried to kill them."

I almost figured out what was going on. Almost. It slipped my mind's grasp before I could stop it. The person she was talking to, the patient in the bed, was somewhere between Dave Steele and Tom Benton.

Tom Benton is the protagonist in a novel. I'm the author. Benton is an American who went to a world called Sexton. He was trained to be a killer/cop in a world of swords and sorcery. Rather than serve in a force he grew to recognize as evil—the protectors guild—he became an outlaw. He took the name 'Viper' and is merciless in his defense of freedom.

She locked eyes with me. "You are a nice man," she said again. I'm not sure if she was trying to convince herself or me. "Every time you started to hurt me, all I had to do was tell you to let me go and you let go right away."

Hurt her? Kill? I was shocked. "I can't hurt anybody. I'm a little guy."

She shook her head, lips pressed together. "You're a strong little guy. Very strong! Don't play weak little man with me. You're strong!"

I knew that—I just didn't want anyone else to know it. Then I saw the brace on her left arm and almost remembered. I didn't remember what happened, but there was no doubt in my mind I was the one who hurt her arm. I was devastated.

I think I was crying. "I didn't mean to hurt anyone."

She tried to calm me. "Of course you didn't, you're a nice man. You were scared, that's all. Absolutely terrified." She stood up and moved closer. "You are forgetting things. You must remember this, even if you remember nothing else. You can never drink again."

"I don't drink." It's true—Viper doesn't. Not much. It would get him killed. She didn't know who she was talking to. She was talking to Tom Benton, AKA Viper. I was gone. Dave? There's no Dave here, man.

"Yes you do!" Her anger was genuine. "You were drunk when they brought you in here! You were almost dead. We saved you this time, but you're not safe yet. We can't do it again. If you drink again, you will die. *Remember this...*you can never drink again."

I didn't remember that. Not for a long time.
Blackout...

7

"Time to go!"

I thought someone shouted that very close to my ear, but when I opened my eyes I was alone. Alone in a bed in a hospital room. I remembered being tired, but I couldn't remember picking a bed and going to sleep. I had to get out of there before someone found me.

I tried to get out of bed, but my left arm was strapped to the rail. *Idiot!* Fell asleep in the open and let the guild tie me down. There wasn't time to wonder why they didn't kill me. The voice I heard in my head might have been imaginary but it wasn't wrong. It was time to go.

There must have been a little of myself left. I looked at the table next to the bed to see if there was a note from someone, maybe Sarah, telling me to stay put or run like hell. When I found no note, I almost called for help.

Viper answered that one. *If you call for help, who do you think will come? Those friendly people who strapped you to this bed, that's who!*

That was all the answer I needed. I stared at the binding. I rubbed my forehead…and laughed. "Stupid guild bastards! You forgot I have two hands."

Now that I had an assessment of the problem, all I had to do was untie myself… I lost a few minutes while I tried to decide what to do after I freed myself. Should I wait there and kill my captors, or leave and recruit an army to come back and kill them?

I decided to leave and let the chips fall as they may.

Untying myself proved more difficult than I thought. The restraint didn't buckle or lace. There were two straps holding it to the bed. I reached over my body and the lower rail with my right hand and followed them down as far as I could reach. The end of the straps was beyond me. I tried another way: tight over my chest, reaching down between the top of the mattress and the bottom rail. I rolled to my left as far as I could.

My hand hit the end of the straps. They were wrapped around the bed frame and looped through a double d-ring. I could picture it—two straps going under both rings, then around and between the curved part of the 'D'.

I'm left-handed. It was more difficult to untie the straps with my right hand than it would have been with my left, but I could do it. In a strange way it was almost refreshing. I found no ambiguity: I was a prisoner against my will; I had no higher priority than freedom. It didn't take long to untie the thing, even right-handed. That's when I noticed the needles in my left arm. They led to IV bags on a stand above my shoulder.

Blackout: A Look Inside Wernickes

The needles worried me. I hate them. I was afraid I would pass out when I pulled them out, but then I chuckled. I was already in bed. If I was going to pass out, what better place to do it?

I was pleasantly surprised to find it doesn't hurt to pull the needles out. It only hurts when they stick you. When the first needle dangled over the floor, I thought about the mess it would make when the fluid dripped out. I laughed at myself again—what kind of prisoner cleans up after himself when he escapes? The empty restraint would tell them I was gone whether there was a mess on the floor or not. I pulled the other needles out one at a time, jumped over the bed rail, and headed out the door.

No one seemed to notice me in the hall. It was daylight and there were staff and patients around. As I approached the door to the stairs I realized I was wearing only a hospital gown. That was fine in the hospital, but would look mighty strange on the streets.

Our buddy Tom Benton is a resourceful guy. The plan was to go down the stairs and find someone of similar build, knock him out, drag him to a closet, and steal his clothes. Then walk home taking back streets.

I was foiled at the door to the stairs. It was locked. The lock was old, a combination lock with small buttons and Roman numerals engraved in the brass above them. I stared at it and tried to remember what little I knew about the type. I didn't think the combinations on them were changed easily, and therefore weren't changed often. That meant the buttons involved in the combination would be more worn than the buttons not involved. Of the seven buttons on the lock, I could narrow the possibilities to three or four. Good idea, right? It would have been if they gave me time.

David J. Steele

I heard a shout from behind. I turned to look over my shoulder at the door I left only a few minutes before. A big redheaded guy came out. He saw me and shouted *"You! What are you doing out of bed?"*

I wasn't going to stand there and take his pop quiz. I turned to run in the other direction, but couldn't. There were patients in wheelchairs by the window. They blocked my way. I thought about jumping over them, but ruled it out as an option. Good guys don't risk hurting the injured and infirm trying to escape. I had to go through the guy I've come to call "Big Red."

I gave myself up. Before I knew what was going on, there were several people around Big Red. He stood behind a wheelchair, waiting for me. I sat in the chair and wondered if they were going to kill me. The other people gathered around us. The thought of springing from the wheelchair and doing as much damage as I could crossed my mind. I discarded the idea. Live to fight again, I thought. Can't win if you're dead.

Much later it occurred to me that the people surrounding the wheelchair—the people who formed a wall between the patients in the hall and the lunatic in the wheelchair—were all women.

They knew I wouldn't hurt the women.
Blackout...

8

I woke up in artificial light, not torchlight or candlelight. It was America, not Sexton. Something went wrong with the crossing this time. I was in a straitjacket in a hospital bed. My arms were crossed over my chest under the jacket. A shake of my shoulders told me I wasn't going to fight my way out. The restraint was tied to the bed at the shoulders. I couldn't reach anything with my arms bound like that.

Blackout: A Look Inside Wernickes

I closed my eyes and tried to remember everything I knew about straitjackets. It didn't take long to figure out I wasn't in one—not a straitjacket, but something else. There was no buckle or strap under my back. I had seen a straitjacket at some point in my life and hadn't forgotten the fear I felt when I saw it. I studied it for that reason—I tend to study things that scare me in case I ever need to play to win.

I concentrated on my legs, particularly around the groin. Straitjackets strap between the legs so the patient can't worm out the bottom. That thing had no such strap. And the guild bastards actually did me a favor by lashing the thing to the bed at the shoulders. If the restraint wasn't strapped to the bed at the top, there would be nothing holding it in place. Holding in place prevents the prisoner from getting out of bed, but that was a two-edged sword. Holding it to the bed gave me something to pull against on my way out the bottom.

I anchored my feet to the bottom of the mattress and pulled myself out with my legs. When I messed up my hair dragging it over the canvas, I wished I had a comb. Not because I worried about what I looked like, but because it would be easy to spot the escapee by his messed-up hair. I shouldn't have worried. Have you seen the typical haircut in Cleveland?

I made it out of the room and found myself lost in a sea of blue-green curtains. It was a house of mirrors with no mirrors, only curtains.

Blackout...

9

I was in a wheelchair looking down a set of metal stairs at a machine with a tube-shaped entrance. A woman stood next to me and was explaining something, but I didn't understand her.

My attention was on the tube. It looked like a tight fit even for a little guy like me.

"I'm claustrophobic," I said. I'm not, really. I just didn't want to go in the tube.

"It's okay." She was trying to reassure me and it worked. I sensed no lie in her face or voice. "If you get scared, just bang on the side. Or say you're scared. We'll let you out."

Sure they would let me out. She was one of those friendly people who strapped my ass to the bed. "You can hear me in there?"

"Yes. We'll be able to see you and hear you."
Blackout...

10

I don't know what went wrong in the MRI chamber. Something did. It wasn't the test, I'm sure of that. Magnetic resonance imaging is a great tool. Expensive as hell—I saw the bill later—but a great tool.

Speaking of Hell...that's where I went next.

I was on my back in the tube. It made me think I was in a Dreamsicle. There were bands of orange and white light, like the vanilla ice cream in those frozen orange treats.

That's what I remember. They tell me I'm wrong, but that's what I remember. Orange bands with beams of white light.
Blackout...

11

I opened my eyes and saw a white ceiling. I was myself at that point and thought I finally got to be the guy in the bed in the bed races. My dad took me to see them when I was a kid and I thought it would be fun to be the guy in the bed, riding right down the middle of the street as we raced to the cheers of the crowd.

The lights in the ceiling were the only clue I had that we were moving. They passed quickly overhead, one every few seconds. Flash, slide-slide, flash, slide-slide, flash, slide. I tried to sit up but a hand pushed me back.

"Lie down," someone said. "Don't move."

The bed race that wasn't a bed race was a lot more fun than the ride on the train that wasn't a train.

Blackout...

12

My pulse speeds up when I think about what happened next. It still scares me. I was on a bed in a big room and there were other beds around mine with no curtains separating them. There were three or four men lying on similar beds. The room was big and had a curved white ceiling like an airplane hangar where the roof and walls are one, stretching from the floor in an arc from side to side.

I heard the men talking to each other in low tones. It seemed like we were waiting for something, but I didn't know what or who. I asked them what we were waiting for. They didn't know. It felt like a scene in *Waiting for Godot*.

"Why don't we get up and look?" I pointed at the curtain separating us from the rest of the room.

"You can get out of bed?" one of them asked. "If you can, you should."

The next thing I knew, I was standing in front of a row of vending machines. I was hungry so I thought I'd grab a bag of chips while we waited for whatever train we were waiting for. The place looked like a gleaming white version of a subway station.

David J. Steele

I reached in my pocket for change and discovered I didn't *have* pockets. Hell, I didn't even have pants! I was wearing a hospital gown and it was chilly in that place. I like pants—they're one of man's greatest inventions. Never underestimate the importance of pants.

Not long after that, I found myself underground on a walkway between two sets of railroad tracks. There were two trains and they looked ready to leave the station. The train on my right was white. It was no cleaner than a normal passenger train. Lines of people of all ages waited to get on. They were quiet and didn't seem to be in a hurry. I didn't think much of it, but because I had no destination in mind and don't like to wait in line if I don't have to, I looked at the train on my left.

It looked older, like something out of the 1950's. Its sides were orange and beat up. The windows were tinted. I couldn't see much through them, only vague shapes of people moving around. I heard laughter and the sound of music—good rock 'n' roll music—pump through the walls. Party train.

I didn't get on one train or the other. My Spidey-sense was tingling.

The next thing I knew, I was outside the station between the two trains. Orange-white halogen lights burned the darkness. It was cold. Frigid, wicked, bite-ass cold.
Blackout...

13

I was hiding behind some equipment, down low and out of sight. I could see the bed I came from. I wasn't sure why I was hiding—was it to escape, or had I set a trap?

A woman walked in, alone. I watched her look at the empty bed and around the room. I could see she wasn't there to hurt me, but she was afraid of something. I hoped it wasn't me.

She saw me and suddenly *I* was afraid. I wasn't afraid of her or anything else I could name. I was just afraid.

She came to me murmuring gently and took me by the hand. Convinced me I belonged in the bed. I climbed up and was on the train again before my head hit the pillow.
Blackout...

14

I sat in a row of three seats. I was on the window side of a passenger railway car. There wasn't anything to see out the window but darkness. I was alone. Not for long. In a blink I was in a bed...but still on the train. There were other people in beds on the same car, but this time no one said a word. I heard the rattle of the train and felt side to side rocking as we made our way wherever we were going.

It was boring, riding like that. I wished I was back in Peru, riding on the roof of the train like I did in 1987.

...Then I was. I still wasn't wearing pants, but I had a wool poncho to wear against the cold. I knelt on top of the train as it cruised down the side of a mountain in darkness.

As long as I'm at it... When all is said and done for me in this world, when it's time to shelve this little body of mine, somebody do me a favor. Please? Open the coffin and make sure I'm wearing pants. I understand the soul leaves the body, but is it too much to ask to be buried with my pants on? Boots are optional. Please, for the love of God...gimme my drawers.

Yea though I walk through the
Valley of the Shadow of Death
I will fear no evil,
For the Lord is my shepherd
And I'm wearing pants!

David J. Steele

> Dave Steele's proposed epitaph
> 15

We were going down the side of a mountain in darkness. Cars stretched ahead as far as I could see. The train bent around the side of the mountain in the distance ahead. I couldn't see the engine. Behind me, up the mountain, I saw only more cars.

The light on the mountain was red. No sun ever lit those mountains. The sky was an umber, orange and black blend of smoky color. Looking at the jagged mountains was like looking at the tip of a cigarette smoked too quickly—gray ash on sharp orange points.

I was in a lifeless land and the journey was only going to get worse when we reached our destination. I wondered if I should jump off and take my chances living off the land. I looked at the charcoal ground passing along the tracks below. Nothing lived there, at least nothing recognizable. If I jumped I might be able to find small animals to eat...but I was pretty sure I wouldn't want to eat them.

Then it struck me that I was dead. And if I was dead, this wasn't what I was expecting. It wasn't what I was promised. This was... *No!* This wasn't Hell. This was on the *way* to Hell. *En route.* I'm going to Hell, I thought. Not tomorrow, not next year, not maybe. *I'm going to Hell...RIGHT NOW!*

I was terrified and pissed off. I knew if I was still on that train when it reached the bottom of the mountain, I would be in Hell forever. And it *pissed me off.*

The first time I remembered this and remembered yelling, I thought I was yelling at God. Now I know it for what it was: the prayer of a believer in fear for his immortal soul.

I turned my face to what passed for sky. "Where are you guys? Where is God? Where is Jesus? I believe in you! I'm human and I fail. I've sinned. I haven't always believed, but I do now. And this is what I get? No hearing? No chance to hear the charges against me? You send me to Hell without a fucking word? I demand a hearing! Dear God, give me a chance to hear the charges against me and let me defend myself!"

I shouted until I ran out of breath. Then I waited. I wondered what I would say if I got the hearing I requested. I didn't have to wait long.

A hole opened in the orange sky. I was on my back, looking up at the face of a woman. Not an angel, though she might as well have been. It was a woman, a living woman, and I was glad to see her. She said something, but not to me. I couldn't understand her words, but I didn't care what she said. I was more relieved to know I wasn't going to Hell. At least not that day.

I wanted to take proof with me. Something to prove there is a literal Hell. Ironically, the knowledge brings comfort of a sort. Seeing Hell did something for me its ruler desperately doesn't want—it proved to me beyond a shadow of a doubt that there is a God and that Jesus is the Christ.

Relax. I'm not going to preach to you. I'm telling you what I saw. This is a report, not an epistle. Now *there's* a word I don't use every day!

16

I lay on a bed on my back, comfortable under a sheet. I heard a human rustle, like someone trying to be quiet and not doing a very good job. I opened my eyes. There was a man in the room with me. He was wearing scrubs but I don't think he was a doctor. I tried to say something to him but nothing came out.

He came close and stared at me for a second, then smiled. I don't remember what he looked like, but I was glad to see him smile. "You're back," he said. He said it as if greeting a visitor he'd hoped to see but wasn't really expecting. "We didn't think you were going to make it."

I have to tell you—when you find yourself back in this land after seeing the place you least want to spend eternity, after praying as you've never prayed before—the last thing you want to hear is "We didn't think you were going to make it."

It would have been much better to say, "It's about time you woke up." Then we could all pretend it was a dream.

It wasn't.

Blackout...

17

A voice disturbed my sleep. I kept my eyes closed, careful not to move while I assessed the situation. My hands were at my sides as I lay on grass in a valley. I moved my right index finger slightly and tried to find my sword. I almost opened my eyes when it wasn't there. It's never far from my side when I sleep. You don't take chances in Sexton.

"What's your name?"

If I tell them I'm Viper, they'll kill me.

I opened my eyes and saw a hospital curtain and a young woman with a clipboard in her hand. She looked at me and said, "What's your name? Do you know where you are? Do you know what day it is?"

"Tom Benton." My voice cracked when I said the name. I was surprised when she got mad.

"That's *not* your name! There *is* no *Tom Benton!* We checked!"

As I slipped back to sleep I thought, I ought to know who I am, lady. If you're so damn sure I'm not Tom Benton, why don't *you* tell *me* who I am instead of wasting my time and yours?

Blackout...

18

"What's your name? Do you know where you are? Do you know what day it is?"

Damn. Here we go again. Don't tell her you're Tom Benton. Make up a new name. I thought for a moment and opened my eyes. Same gal, same clipboard. Different answer. "David Steele." I liked the name, but it didn't sound right.

She smiled. "That's right."

Right? I'm the *author?* "Author" rang through my mind as if it meant some kind of god. Not with a capital 'G' but a little 'g.' Creator (small 'c' intended, but sentence structure doesn't allow) of the world of Sexton.

"Do you know where you are?"

"Hillcrest Hospital." I thought I was making up the answer. Actually, I *was* making it up.

"That's right!" Her smile could have lit Manhattan. "Do you know what day it is?"

Feeling confident, I said, "July first."

Her face changed just enough to let me know I missed. She forced a smile. "Well... Two out of three is good."

"Which one did I miss?"

She thought for a second, debating whether to answer my question or not. "It's July 5th."

Blackout...

19

There was someone near me, closer than usual. I wasn't afraid when I opened my eyes.

A pretty girl was next to the bed. She had brown hair and smiled with lips I wanted to kiss. She was dressed like a civilian, but I don't remember what she was wearing. I looked at her face; I could have looked at it for a long time. She talked to me, but I couldn't understand what she was saying. It didn't matter. If I can be with her, I thought, everything will be alright. We can face anything. I saw she was crying and wondered why.

I was getting sleepy again. She kissed me! Lightly and lovingly. I was smitten with her. Guilt hit me like a wave. *You're married to a wonderful woman! Don't fall in love with this one!* I turned my head away from the pretty girl. I love my wife very much, and as wonderful as the pretty girl was, I won't cheat on my wife. I knew I was fading again. My last thought was…

…Why does the pretty girl cry?
Blackout…

20

Later—almost a week later—as I got ready for bed in the recuperative care facility, I wondered again, *why does the pretty girl cry?* The answer hit me. It's a good thing I was alone in the room then because I thought it through out loud. Well, half out loud.

"Why does the pretty girl cry?"
Because her husband was lying half dead in the ICU.

Blackout: A Look Inside Wernickes

"Sarah was the pretty girl? ...Of course she was!" I laughed and cried at the same time. Who else could make me feel that way? You can't cheat *on* your wife *with* your wife.

I went to sleep happy.

David J. Steele

21

I opened my eyes. The ceiling was different: squares with holes in them rather than the usual rectangular tiles. I was resting comfortably on a bed, no restraints on my arms or chest. The curtains around the bed were white, not blue-green. I heard two women talking somewhere close.

"Dave Steele is your husband?" one asked.

"Yes."

"I think I met him somewhere before."

"He works for the Boy Scouts."

The first woman said something nice about me, but I don't remember what it was. It didn't matter. The other woman's voice was Sarah's! I climbed out of bed over the railing and staggered when I hit the floor. The landing was good but my balance wasn't. I stood at the foot of the bed, head hung low and whispered to myself. *"Dangerous. They said I'm dangerous. Oh God...I didn't kill anyone, did I?"*

I was sure I would remember if I had. I thought I might black out again and I didn't want that to happen. I told myself to stay in control, remember who I am and stay in control. I sucked in my breath and clenched the metal rail. "You're not in Sexton. This is America. Kill in Sexton when you have to, but never here. Never kill in America."

I turned and peeked through the curtain. I saw Sarah's back. She was pressed against a counter, talking to the woman behind it. She was wearing shorts and a t-shirt. I checked out her legs...always loved 'em. Then, like a good little patient for a change, I crawled back in bed to wait.

Blackout...

22

I was in a wheelchair being pushed out the front door of the hospital. It was a beautiful summer day. Sarah's car was waiting. I waved to people I didn't recognize because it seemed like they were saying goodbye. Whether they were happy to get me the hell out of there or happy I was better, I can't say. I hope it was the latter and not the former.

Big Red was there. He was smiling, and for the first time I didn't see him as an enemy. I think he was a little sad to see me go. I don't know why I think that, but I do. Someday I'll meet him and learn his name. I'll shake his hand and thank him. I owe him an apology. I haven't written the scene in here—I don't know where it fits in the time line. Frankly, I'm not sure where anything fits in the time line. I've assembled this in the order that makes sense to me now.

Big red came to see me when I was conscious. He told me that if I kept getting out of bed I was going to get hurt. That didn't faze me. Then he told me that if I got out of bed someone *else* would get hurt. I took that the wrong way. I know now that he meant that *I* might hurt someone. I thought he was threatening Sarah and told him that if anything—*anything*—ever happened to her, I would destroy his life by taking out everyone who ever meant anything to him and we would go to Hell together with me on his back.

I won't apologize for the threat. That's what I would do if someone were to harm Sarah. I apologize to Big Red for not trusting him when he tried to help me.

David J. Steele

Part II
Which Green?

1

I was surprised when Sarah got me home. It wasn't the house I expected. I expected the gray and pink colonial we owned in Chicago. This house was smaller and older, but looked pretty cool.

It seemed like a nice house. I liked the furnishings and was very interested in the china cabinet. It made me feel comfortable. I was sure there were stories behind some of the nick-knacks in there. I heard a noise down the hall and wandered toward it. I found Sarah in a bedroom putting things in a bag. I don't remember what she said or what I said. I wanted to stay and couldn't—that was the bottom line. I'm pretty sure we were both crying when we left the house to go...wherever it was I had to go.
Blackout...

2

I was in a room. It wasn't a hospital room, but it resembled one. I was just sitting there looking at the night through the windows. I didn't know enough to be either lonely or afraid. For all I knew, I'd been there forever. Sitting in front of the dark windows wasn't the culmination of events. It was all there was.

"Mr. Steele, you have a phone call. I think it's your boss."

I was led down the hall to a high counter. There was an empty chair behind it. Someone left a phone on the counter for me. The handset was off the hook. "Hello?"

"Dave! It's Mike Stone." The voice sounded vaguely familiar and it made me smile. "How are you man? We were worried about you! You okay? Are you better?"

"Sure. I'm fine. How are you?"

We didn't talk long. I'm pretty sure that conversation wasn't my best. I hung up and walked back down the hall to my dark windows. The personnel in that place did a good job pretending to give me privacy.

"Was it a good call?" a woman asked from somewhere behind me.

"I think so."

"Who was it?"

"Some guy." I didn't turn around to look at her. "Seemed like a nice guy."

"Do you know who he is?"

"No."

I had trouble falling asleep that night. There was a bed in the room. It was comfortable enough, but I tossed and turned for a while. Finally the light came on and a man said, "It's time for your meds. Take these."

Blackout...

3

"*Wake up!*" a female voice shouted. "Breakfast will be here in ten minutes!"

I opened my eyes. I was on my back in a hospital bed. The room was different than any hospital I remembered. It was bigger and the curtains between the beds were pulled back. There were windows on my right. I was on the ground floor. Outside it was sunny and bright. I could see blue sky over the other side of the building—evidently it wrapped around a courtyard.

David J. Steele

There was a bed on the opposite side of the room. The patient in the bed was an Asian man of late middle age. He didn't acknowledge me and I didn't blame him. He was eating. I heard a woman tell him he had to finish everything on his plate or he would be in trouble. That worried me. I almost never ate everything on my plate as an adult. I was never as hungry as most people.

I didn't have to worry. The food was hot and good. I felt like I hadn't eaten in a week. As I shoveled sausage and scrambled eggs into my mouth, I tried to remember the last time I ate. I couldn't.

When I finished the meal I felt the tugging of a nicotine fit. I was wearing shorts and a t-shirt. I felt the pockets for my cigarettes and lighter. They weren't there, but I saw them on the windowsill. I couldn't remember the last time I had a cigarette, and I wanted one.

The woman who took my plate told me where I could go to smoke. She helped me out of bed and into a wheelchair, then she gave me the rules: no walking anywhere—not even to the bathroom—without permission from the doctors; the smoking area was in the courtyard; lunch would be served in my room, and if I wanted to eat, I'd better be on time.

I rolled down the hall, around a corner and out a door. Sunlight in the courtyard, wheelchair under me. Pulled out the pack of Marlboro Lights, lit one, sucked in the smoke and let it out slowly. The cigarette, like the breakfast, was good and gone before I knew it.

4

The courtyard was plain. The building bordered it on three sides. One side was the wing I came from. My room was on the end. The opposite side held the door we used to go in and out and a common room with couches, tables, chairs, and vending machines. The other side was a glass enclosed walkway. The open end was a nice grassy rectangle with houses on the other side of privacy fences and hedges.

It was a nice place to sit and smoke. There was a cooler by the door, a plastic thermal cooler with a white spout at the bottom. The water was always cold, even in the afternoon. There were people in the courtyard. Some were in wheelchairs, others used walkers, and still others seemed to need no assistance.

A grizzled old guy in a wheelchair pulled himself toward me, making eye contact every inch of the way. Something about him seemed a bit off…I couldn't tell what it was, but I paid attention. He wore a dark blue ball cap on his head with a U.S. Navy ship's name on it, but I don't remember which ship. His language skills were minimal, but I had no doubt what he wanted. He wanted a cigarette. I know the look.

I was about to give him one with a smile on my face, but a voice stopped me. "Don't give him one, man! He always wants one but never has one for anyone else." Before I turned my head to see who spoke, the old man growled something at the guy and gave him the finger.

I never broke eye contact with Captain Ahab. "I'm out," I said.

The guy in the wheelchair—I call him Captain Ahab and don't care if anyone else does—pointed at the pack of cigarettes next to my right hip. Pointed at the cigarettes, pointed at his mouth, then back at the cigarettes.

I got the point, smiled, and shook my head.

He half grunted, half growled at me. As if to make his point, he swung his fists in the air. Then he held one up like

David J. Steele

Jackie Gleason in the *Honeymooners. Bang-zoom, Alice! To the moon!*

    I didn't laugh at him. For all I knew he was tough as nails and would kill me for a cigarette, but I knew he would never get out of the wheelchair. *Could* never get out of the wheelchair. I stood up. I could get out of my wheelchair. I wasn't sure I could walk…but I could get out of the chair.

    Now that I was standing, the guy in the wheelchair decided he didn't want a cigarette after all. At least not from me. He skulked back to his corner. I turned to look at the guy who told me not to give the old man a cigarette and found myself looking at a tall black man, about sixty years old. He had a warm smile and was wearing a Panama-style hat and Hawaiian shirt, and was sitting on a white plastic chair with a walker and a standing ashtray at his side.

    I said, "I'm going to get a drink of water. Want one?"

    He did. That's how I met my friend, Bill. People can say what they want about Bill, but he became my friend that day. I needed one in that place, and he was it.

5

I thought too much that day. By the time I went to bed, I was determined to get out of there. I was pretty sure the place was a nursing home and as far as I could see I was the youngest patient. Thirty-nine years old was too young to be stuck in a nursing home. An orderly came and gave me pills. I tried to hide them in my mouth, but he got me to swallow them. Not through coercion or violence—he just stood there and waited.

    It was deep in the night when I forced myself awake. I sat up in bed and tried to blink the cobwebs out of my eyes and skull. The man in the bed on the other side of the room was out cold. I heard him breathing deeply. Dim light filtered in from the hallway and I could see his shape in the bed on the other side of the room. The only other light in the room came from the stars outside the window and the night light in the bathroom.

I walked by the wheelchair waiting for me at the foot of the bed. My gait was unsteady, but I could walk. I was me again—Dave Steele, not Viper—and I was getting the hell out of there. In the hallway, I turned right instead of left. There was an exit at that end of the hall. I was pretty sure there were no security monitors in the hall—I checked when I came in from my last cigarette.

I had to lean against the wall as I studied the exit—an alarmed emergency exit. I was still groggy from the pills.

The plan in my head was vague: walk home and talk Sarah into relocating to South America...hide from whatever force was keeping me prisoner. *Invoke Plan Z.* Plan Z is for use only when Plans A through Y fall through. I won't give you the details—there aren't any—but it doesn't involve destroying anything. Just a very long walk to a remote town in Peru. You'll know I've invoked it if you ever find a note on my desk after all is lost. The note will say, "Gone to Hot Water." If anyone figures out what that means, I'll buy them a Cuba Libra at the Inca baths, but won't have one myself. Look for an Irishman named Mooch O'Grassyass. That'll be me.

I stared at the door for what felt like a long time. There was an alarm hard wired to the building. I couldn't think of a way to disable it without setting it off.

This time the voice in my head wasn't a fake. It was my voice and represented the clearest thought I had in a long while: *What makes you think you can convince Sarah to go with you to Peru? Don't you think she would move Heaven and earth to get you out of here if she didn't think you should be here? She loves you! Trust her if you trust no one else. Go back to bed.*

I listened to myself. I'd keep track of the days, and if I was still there thirty days from that day, I would break free and walk south.

6

Bill and I didn't do much together. There wasn't much to do but sit in the sun and smoke. Sometimes we talked, and sometimes we didn't. We went to our rooms for meals and came back to the courtyard when we were ready. I don't remember how it came up, but he said something about his evening beer. How nice it was to end the day with a couple of beers.

Honest to God, I had no idea why I was there. When Bill said his granddaughter brought him beer and told me I should ask if I could have some brought to me, I planned to do it. I would have asked, but something stopped me. It didn't seem like the right thing to do.

A tingling feeling that something was wrong didn't stop me when Bill came to see me one afternoon and told me to follow him. He had a white plastic grocery bag under one arm and led me to the courtyard. There was a pavilion on a concrete slab. It was a hot afternoon and Bill and I were the only people outside. He snuck two cans of Miller Genuine Draft from the bag and slid one over the picnic table to me. "Drink up, quick!"

I wondered why two grown men had to hide beer in the middle of a hot summer day. We were both well over twenty-one years old. If he was allowed to drink in his room, why couldn't we drink outside? I leaned back in my wheelchair and took a long, slow drink. It tasted fine. Better than fine.

He started his second beer by the time I put the can down from my first few swallows. Something about his demeanor made me feel like I was an eighteen-year-old freshman on my alma mater's dry campus. I saw the curtain in the break room twitch. Someone saw us drinking the beer, but didn't want us to know they were watching. It piqued my curiosity—the spying—but we weren't doing anything wrong.

Bill slugged back his second beer and said, "Drink up, man! Come on!"

I like beer... No question there, but I never liked to hurry through it. I was lightheaded and slightly ill. It had been a long time since one beer did that to me. I took a couple of sips from the second can, but then I had to stop. My head was spinning.

I wheeled myself back into the building, through the door and down the hall, weaving as well as I could through a forest of people. When I got back to my bed, I hauled myself out of the wheelchair and slept until dinner. It was a black, dreamless slumber...

7

There was a storm. It was mid-afternoon and I was sitting on the bed, just finished with lunch. Clouds rolled in thick and dark over the courtyard. I watched the leaves blow crazily on the trees. The sky had the pale amber look of a severe storm, the kind that precedes a tornado. Wind whipped the air as if cued by my thoughts. Silver drops the size of quarters beat against the windows. Lightning flashed. The lights dimmed. Then the storm was gone as quickly as it rolled in.

I got in the wheelchair and was about to go out for a cigarette when I saw Bill. He stood in the doorway to my room, panting as if he'd just finished a marathon. He was wearing a nylon practice jersey. God help me, please don't ever make me look at another middle aged man's nipples!

He stuttered sometimes, but that time it was bad. "There's...the...there's..."

"Slow down, Bill. What is it? Take a breath."

"...*Big dog!* Wa...*woulda*..."

"Where?" I looked out the window. The sun was shining through the remaining clouds. There was no dog in the courtyard. Not that I could see.

"Out there. It came from behind a fence. Big dog!"

I told Bill I was going out for a cigarette and wasn't going to let a dog stop me. Not a big dog, not a little dog. I'd take care of it if it was still out there. I'm glad he didn't ask me how I was going to take care of it. By the look in his eyes—a little fear, a little respect, and more than a little trust—I could see he didn't want to know.

I wasn't going to hurt it, but I'm pretty sure the animal rights activists wouldn't approve of my method if I needed to drive it off. The plan running through my head was to hop out of my wheelchair and chase it off with Bill's walker. Eyewitnesses wouldn't have sufficient credibility. Who would believe their story that the little guy jumped out of his wheelchair and ran off the dog, swinging the walker owned by the big guy on the ground?

8

I think the physical therapy sessions started the day I was allowed to eat breakfast in the dining room. I say I think because time did strange things to me—okay, my *perception* of time did strange things to me in that place. Someone told me I had an appointment for physical therapy at 1:00 PM that day, where in the building to go, and told me to be on time.

Blackout: A Look Inside Wernickes

I was getting annoyed at people asking me to be on time. I'm punctual to an extreme: never more than three minutes early, and if your watch says I'm a minute late, it's probably fast. It could have stopped eleven hours and fifty-nine minutes ago. I'm open to possibilities.

Breakfast was good. I rolled to the dining room and took a seat with three other wheelchair bound men. There was a menu —half a sheet of paper with choices for eggs, cereal, fruit, toast, and juice. We were encouraged to order something from each group. Someone brought coffee. I circled my choices and the food came quickly.

At precisely 1:00 (I waited for five minutes around the corner in the hall,) I rolled into the physical therapy room. It was about the size of two patient rooms combined. As I came in the door, I saw a computer attached to some kind of gizmo that helped patients with their grip. An older woman sat in front of it, squeezing a handle at random intervals. To my right was a set of parallel bars. In front of the door, about seven feet away were wrestling mats. A man in sweats was lying on one. The therapist was on the floor next to him, encouraging him to squeeze an inflatable ball between his thighs as hard as he could. I tried not to think about why a man needed to be able to squeeze hard with his thighs... It's probably best if you don't either.

The therapist smiled up at me. I don't remember all the exercises he had me do, but he wanted me to concentrate on a set of four wood stairs in the corner. He said my ability to use stairs was important because I lived in a split-level house. The term brought back a memory—just those two little words. We didn't live in the house at 1326 Townsend anymore. We lived in a split-level house at...somewhere in...Cleveland? Yeah—Cleveland.

I climbed the stairs easily in spite of having been in a wheelchair for a while. In fact, I got a little bored going up and down them with and without using the banister. Then I got tired. Fatigue hit me like a wave. I thought age was catching up with me. Not long ago I could have taken stairs two at a time without much thought. There was no way I could have done it that day.

At the end of the session, the therapist gave me permission to walk. There was caveat—I was to keep the wheelchair with me at all times. He said to push it in front of me and sit if I got tired.

First breakfast in the dining room, then walking behind the wheelchair. Considering I still didn't know I was sick, (I know, how could he *not* know?) it felt like things were looking up.

9

Dad and Sarah were my first visitors. That morning, one of the staff told me I would have visitors. I didn't have to be in my room; someone would tell them where to find me.

I was sitting in the morning sun, smoking. I'm not sure who saw who first, but I was very glad to see them. We had a nice chat. I don't remember the topics—it was just a pleasant afternoon. Loving hugs were exchanged, then they left.

I was in a great mood for the rest of the day. I couldn't remember why…

10

The occupational therapist was a small woman. Dark hair and a bright smile. She was all business.

The session didn't last long. She walked behind me to the bathroom—her idea, not mine—and held me with a strap around my chest in case I fell. I could have saved her the time and worry. I knew damn well I could walk to the bathroom and take care of things without a babysitter. Thank God she didn't need proof of that. She wanted to see if I could walk, unbutton my shirt, put on my watch, etc. I could and did.

Questions formed in my mind. The nightmares that disturbed my sleep invoked hospital images too many times. Had I been in one? Was I in a coma? Why the hell couldn't I remember things I knew happened? Pieces and feelings, clips and snippets, were all I had.

She and I sat near the counter in the hall. I asked her if I was in a coma and her answer surprised me. "We don't call it that anymore," she said. "It freaks out the family."

Highly technical medical term, freaks out. I'm not qualified to use it. "So I wasn't in a coma?"

"We prefer to call it a non-responsive state."

"Sounds like a coma to me."

She shook her head and smiled. "Non-responsive."

You can call a rose a cheeseburger if you want, but it still won't taste good with a pickle. Non-responsive. She was pissing on my boots and trying to tell me it was raining.

She asked me how much I drank and I told her the truth. Yes, the actual truth. It didn't surprise her. She said I might have to go to another facility when I was done in the current one—a rehabilitation facility. Given my progress, she didn't think I would have to do that and would recommend against it. I would, however, need physical therapy. My insurance would pay for in-home visits.

Then she asked if I wanted to go on long-term disability. At first I didn't understand the question. Why would I want to do that? I was physically able to work...still had all my fingers and toes. Heck, I could walk, and pee, and everything. I asked her why I would go on disability.

David J. Steele

"Most people who suffer what you suffer aren't able to go back to work." She leaned closer and said, "When you go back to work if you find it too difficult or can't function like you used to, *don't quit*. I can't do anything to help you if you quit."

That conversation went to the dead memory bank until roughly thirty days after my resignation took effect. Sometimes I regret not taking disability, but not for long. Disability would have killed me. It would have meant I gave up. Sure, there would be more money than I make now, but I don't know that I would have worked to bring my mind back if I collected a disability check twice a month. It was no easy task to bring my mind back.

Sometimes it's hard not to regret forsaking disability. Those moments tend to come when I'm scraping stuff—let's call it stuff—from a toilet seat at four o'clock in the morning. The moment doesn't last long. I flush the regret and the *stuff* down the toilet and move on. In fact, I often smile when I do it; life is better now, post Wernickes.

11

The best part of the day was the last part of the day. Bill and I would sit outside with our backs to the walkway between wings, looking out at the courtyard and the neighbor's fences. Nurses came out after dark to wait for the city bus to take them home. I stayed there as late as I could. There was comfort in talking to the ladies…the comfort of knowing I was a person, not just a patient. And knowing *they* were people—regular people—as well as caregivers.

There was a black woman, probably in her late fifties, who sat with us sometimes. I knew her to be a sweet lady. A *tough* lady, but very sweet. I'm not sure what her title was, but there was no doubt she was in charge of the place at night.

Blackout: A Look Inside Wernickes

I sat out there with the nurses every night, but only once did I miss the call for lights out. The call for lights out wasn't really a call unless you were in the building. Apparently I missed the memo that said we were supposed to be in our beds by 11:00 PM for our medication. I was alone when I finished my last cigarette for the day, and the halls were still lit when I wheeled myself to my room. Unlike the previous nights, no one came to give me my meds.

I was afraid I wouldn't be able to sleep without them, so I headed to the nurses station. That's the one to the *left* outside my door. The Head Nurse—my smoking buddy lady from outside—was in the hall, about to lock a cart in a closet. I asked if I could have my meds.

"You missed lights out." She shook her head. "You'll have to go to bed without them."

"I didn't know there was a lights out."

"No meds after lights out."

I recognized the finality in her voice, and I wasn't going to argue with her. I wasn't trying to manipulate her with what I said next, but now that I've thought about it, I don't think I could have played it better if I tried.

"That's okay," I said as I turned my wheelchair toward my room. "This seems like a big place. I'm not even sure where the kitchen is. The food is good, but I'd like to see the kitchen. Take a little tour." I looked up at the ceiling and around the walls as if seeing them for the first time. "Don't worry. I won't touch anything if I don't know what it is, and I promise I won't bother the other patients. I'll explore quietly."

"Mr. Steele, come back here."

I turned. Her tone of voice didn't give me much choice. She looked like she couldn't decide whether to laugh or yank me out of the wheelchair and put me over her knee. I have that effect on people sometimes—I don't know why, but I'll confess that I enjoy it.

"I'll give you your pills," she said, "but next time you have to be in bed at 11:00."

I didn't mean to do it, but I think I gave her what my wife calls my "cute little boy" smile. She couldn't seem to help herself and thawed her face.

She gave me the medication. "Promise me you won't go exploring."

"I promise." I kept the promise. No exploring for me. Maybe that's why I got ice cream...

12

One night as we sat in the dark I said I missed ice cream. I wasn't asking for it; just feeling nostalgic. One of the nurses promised to bring some to my room. It would be there when I went in for the night.

I was surprised when I got back to the room. The table was pulled over the bed, and on it were two little white cups of vanilla ice cream and one of those flat wood spoons, the kind with points on both ends. I laughed like a little kid and slid in bed to enjoy my treat. It ceased to be ice cream sometime before I got to it, but I didn't care. I peeled back the paper lid and stuck the plank in the goo. There might have been one clump in the puddle. The rest was a sticky, white, slightly cool, messy, sweet substance.

13

They told me I would have a visitor that day. Mike Stone—my boss. He's a good man.

I was sitting in the courtyard, smoking a cigarette. Bill wasn't there. I was parked in my wheelchair in the shade. I saw Mike walk through the walkway between the wings. A nurse pointed me out to him. I waved, glad to see him.

I don't remember much of our conversation. He was on his way to camp. He looked great in his white shirt and tie. He wore the tie loose in anticipation of changing into his uniform. We chatted about small things. I told him I planned to leave the Boy Scouts.

If he hadn't left me magazines, I might not have remembered his visit. That's terrible to say, I know. But my memory was patchy at best. He left me two, I think it was two, magazines.

After Mike left, I sat in the room. I wanted to read one of the articles in the magazine. Ironically, it was about some aspect of men's health. I struggled to read it—the words weren't hard, but reading wasn't automatic. I had to think through what each word meant as I read it. I set it aside when they brought dinner. Afterward, I went out to smoke. The magazines were no longer important; forgotten completely until I found them with my stuff after they let me out.

14

The old guy in the bed across from mine finally spoke. I came in from my last cigarette of the night and he was sitting in a wheelchair at the foot of his bed. A restraint tied him to the bed by one arm. I don't know why he was restrained. I had a vague recollection of him moaning the night before. Staff came and took him away.

Now I looked at him strapped to the foot of his bed. His face was slack, almost tortured. His thin hair barely covered his shiny scalp. He looked at me and asked for help, raising his arm as far as the binding would allow. "They forgot me. Please help me… I want to go to bed."

I thought it highly unlikely they forgot him. "Do you want me to call for help?"

He shook his head. "They won't come. It's too late at night. They left me here. Please help me."

I'm pretty sure there's a rule against one patient untying another patient when the caregivers find it in their hearts to leave a poor old man lashed to his bed. I would have left him there if and only if: 1) they bothered to tell me why he was tied to the bed, and 2) the answer made sense to me. They didn't tell me why, therefore there was nothing to make sense *of.* That entitled me to do what I felt was right.

I said, "I'll untie you, then help you get in bed. Stay there. If you need to get out of bed, wake me up or call for help." He nodded, but that wasn't good enough for me. "Promise me."

"I promise." He made eye contact and gave me a weak smile.

I untied him and lifted him into bed. He felt frail under my hands. Once in bed, he leaned back on the pillow and sighed. The exhausted smile on his face warmed my heart.

"Thank you," he said. It came out a whisper. He asked for his glasses, then picked up a book from the table and pointed to the cover. I didn't understand everything he said, but he was trying to tell me about the book. It was some sort of Christian book. I don't remember the title.

In the way of a true evangelist, he demonstrated a complete inability to recognize he was preaching to the choir. The older I get the more that annoys me. I smile as I write this, but I have to admit I felt the very human urge one Christian feels when another Christian tries to convert him to what he already knows. I wanted to haul his scrawny ass off the mattress and strap him to the frame again. Don't shake your head at me—you know what I'm talking about.

"I know Jesus is the Christ," I said. "I accepted Him and He me." The memory of the red mountains came back and I tried to suppress a shudder. "Go to sleep. He'll watch over you."

Something got through to him. I don't know if it was my voice or my words, or the lateness of the hour. He nodded and smiled over his thick glasses. I think he was asleep before I crossed the room and crawled in bed for the night.

15

My nephew Timmy and the Green Goblin both had a hand in saving me and neither one was there. Timmy is my brother's son—a sweet little blond boy, about three years old at the time.

I was sitting on my bed in the middle of the afternoon. One of the things Mom and Sarah brought me was a small black and white television. The news was on—someone the police thought was a terrorist had been shot in an airport. In other news, there was a subway bombing. I didn't know where the shooting took place or which subway was bombed. The outside world seemed like a terrible place. I wasn't sure I wanted to go back.

I was half listening to the news and coloring with crayons. The picture was of the Green Goblin of Spiderman fame flying his scooter-thing near the edge of a skyscraper. When I was a kid, I was terrible with coloring books. I couldn't stay in the lines to save my life, and although I knew what color things were supposed to be, I didn't always color them that way. Now that I think about it, I think we lose something when we stop coloring the way we want and start coloring according to what we see. Maybe the three-year-old has something when the grass is red and the sky is green. Yeah, I know—I just sent forty-three kindergarten teachers screaming out of the room. It's okay, though. They probably needed a time out.

I thought of Timmy as I colored the Green Goblin's armor with the magenta crayon. I could picture us coloring together on a sunny afternoon. I had very little to worry about. Soon someone would come with food on a tray. I would eat dinner and go out to smoke with Bill until darkness fell and we swapped stories with the nurses.

I was about to glide to an epiphany. Self-awareness is the key to sentience. If you choose to say it more poetically, you might say it as Descartes did: *Cogito ergo sum.* I think, therefore I am. I've played with the phrase a bit. I think the more accurate way to say it is *Cogito, cogito, ergo sum.* Translation: I think, I think, therefore I think I am. Or try it this way—perhaps most accurate of all: *Cogito cogito ergo cogito sum…cogito.* In English: I *think* I think, therefore I think I am…I think. Play with the phrasing long enough and you'll get dizzy. It's a little buzz I call recreational philosophy. There's probably a 12-step program…

The Green Goblin's armor looked okay. The magenta wasn't dark enough, but it conveyed the idea. The skin threw me off. My box of Crayola crayons had several shades of green. I had an internal debate with myself. I think it was with me, maybe not. I didn't say anything out loud; it wasn't necessary.

What color is the Green Goblin?
Green.
Which green?
I don't know.
Which green?
I don't know.
Who says it has to be green?

Blackout: A Look Inside Wernickes

It's the Green Goblin.
Is it?
Yes. An enemy of Spiderman. One of many. The Green Goblin.
Who says it has to be green?
Mr. Stan Lee of Marvel Comics.
The Marvel Comics of which you own 400 worthless shares? The Marvel Comics that was bought out by ToyBiz...of which you own no valuable shares?
Ouch. Got me there. Who says the Green Goblin has to be green?
You, tell me
I say the Green Goblin has to be green. It's my coloring book.
Which Green?
Whichever green I choose, ya bastard.
That ended the discussion. Me, myself and I—a committee of one—decided that the Green Goblin is *green*. I no longer cared what color the Green Goblin was in film or comic. I cared only about the color green I chose. I selected a light shade of green, one that seemed more powerful against the magenta background than the shade Mr. Stan Lee and his bankrupt editors chose.

That was a crossing moment for me. As pleasant as it was to sit on a bed coloring in a book and chatting with a nephew and comic book villain who weren't there, I had to pull my mind together or I would remain a man rolling around in a wheelchair, ignoring news of subway bombings and airport shootings while I ate meals from an unseen kitchen and sucked down melted cups of vanilla ice cream.

This time it was *my* voice.
"...Time to go."

David J. Steele

16

It was sometime between Thanksgiving and Christmas when I remembered the intervention. By that time, the desired results were already in play. I'm sober.

They wheeled me into a small conference room. There was a table and several chairs around it. A window opposite the door let in the sun. Sarah and my mother were there with the physical therapist and a couple other people I had seen before but didn't know. They pushed my chair next to Sarah. The woman running the show was on my left, at the head of the table. Mom was across the table from me. There was some discussion, perhaps explanation. Details are sketchy. Someone asked me how much I drank.

"Four to five beers a night," I said. That was a lie, and not a lie.

When I was in Chicago, my boss recognized my drinking problem. He sent me through our employee assistance program to an alcohol education program. They did an assessment and put me in a ten-week class. As part of the assessment, they ask you about your drinking habits, etc. I answered the questions honestly and accepted the help willingly—allowing I might need it, but never believing it. It took having my brain explode to make me see the light.

In my research of drug and alcohol counseling—let me back up a bit. When I hear something that doesn't make sense to me, I research it. I want to know what makes it tick, what causes what, and how to deal with whatever it is. I found that substance abuse counselors are trained to ask the addict about his consumption habits and then *multiply by two*. When the counselor in my assessment asked me how much I drank, I told him the truth. He then multiplied by two because addicts lie…which meant he thought I drank sixteen to twenty beers a night. The real answer was eight to ten.

"When I said four to five, Sarah said, "Stop *lying!* It's eight to ten beers a night!"

It wasn't a long meeting. I listened to what the people in the room had to say, and they were right. I felt loved—exactly as intended. The people in the room were there because they love me and wanted to bring me back from the hell I put around them and me. I'm grateful for that.

Then one of the people around the table—I think it was the physical therapist—wanted to know where I got the beer I consumed on the premises. At first I thought it was a trap to break my spirit by getting me to betray a friend. That thought came from Viper. He was wrong.

I realized they didn't know where I got the beer. *That* pissed me off. If part of their job was to make sure I healed and part of that healing was abstinence from alcohol, there was no way I should have been able to get beer in my hands at all—let alone drink it. I had no memory of what put me there. I didn't know I couldn't drink, but *they* knew and didn't or couldn't stop me. Now they wanted to know which patient gave me beer right under their noses? I wasn't gong to tell them. We could sit there for days, and I wasn't going to tell them.

Mom surprised me. "He's not going to tell you. I can see it in his eyes."

I don't remember the rest of that meeting. Meeting, intervention, whatever you want to call it. Soon it was over. Mom gave me a hug and told me she loves me. Sarah did the same. I was sad when they left, feeling small and guilty. Less than an hour later I was my old self. The intervention was gone from my mind. Blacked out.

David J. Steele

17

Mom came to pick me up the next day. I was surprised and delighted to see her, and even more delighted to find out I was going home. She had some paperwork to do, and I went to my room to pack my stuff: the cell phone, the black and white television, my clothes, and my coloring book.

There was a guy in a white lab coat standing outside my door. I thought he was an orderly, but he might have been a doctor. He gave me a couple cardboard sheets with Propranolol—the medication I took for essential tremor—in plastic bubbles. I asked about the sleeping pills.

He looked confused. "Sleeping pills?"

"The little yellow ones."

"Oh, those. You can't have them when you're not here. It's a narcotic. You have to stop taking it before you get addicted."

I could have slugged him. I thought about dragging him into the room and beating him to death with one of the chairs. What kind of sick bastard gives a drunk narcotics? But I didn't beat him, or even yell at him. "Can I take melatonin?"

He smiled. "Sure. Melatonin won't hurt you." He gave me a discharge form to sign. "The other thing you need is thiamin—Vitamin B[1]. Take it every day. You should be able to get it at the same place you buy melatonin. If you can't find thiamin by itself, take a multivitamin. Just make sure it has thiamin."

"How long do I have to take it?" I forgot the answer until much later. I won't forget it again.

"For the rest of your life. Forever."

I threw my stuff in a bag and walked out the front door. Mom and Bill were waiting for me. It was a beautiful morning. Not early, but still cool before the heat of the day. Mom was driving my car. Her husband Tony was at a hotel, and Sarah was too sick to leave the house.

Blackout: A Look Inside Wernickes

Bill stood by the car, waiting to say goodbye. He looked sad to see me go, but happy I *could* go...I think. The big guy actually hugged me. I'm not sure I would have made it through the days in that place without him. I'm glad I didn't have to try.

Soon I was home. This time I recognized the place.

Bookmarks and Bar Mops

It's been two years since the Wernickes and life is good. We moved to Sarah's hometown, a great small town in the thumb of Michigan. She teaches in the middle school she attended, and I'm writing. I haven't sold anything yet, but I will. I was lucky to survive Wernicke-Korsakoff Syndrome. I had no idea how lucky, until I did some research. Wernickes is an alcoholic's disease...you won't get it otherwise unless you happened to be starving to death, and then only maybe.

I found a few statistics regarding Wernickes:
- 1 in 10-12 alcoholics will get it.
- Of those who get it, 1 out of 5 will die.
- 3 of the 4 who don't die will have permanent brain damage.

I found out the hard way what happens if I stop taking thiamin. There's no reason not to take it. Vitamin B^1 is inexpensive and available at most drug stores. The pills contain 6,667% of the daily-recommended dosage. Sounds like overkill to me, but apparently there's no such thing as too much of that good thing. I mentioned I forgot what the man said when I asked how long I had to take it. Shortly after the first of the year, I stopped. I'm not cheap, but I didn't see the sense in paying money for a bottle of pills that has no effect on anything.

David J. Steele

One day in March, with the sun shining and a hint of spring in the air, I was alone. Being alone never bothered me before and there was no reason for it to bother me that day, but something clicked the wrong way in my mind in the middle of the afternoon. I crossed the living room to go out for a cigarette. Fear hit me like a cold kick to the chest. My pulse hammered in my ears and my mouth was dry. I felt like I bit a piece of aluminum foil. There was nothing to fear, but I was scared to death. I clocked my pulse at two hundred forty-five beats a minute. Cold sweat beaded on my forehead, the back of my neck, and under my arms.

I forced myself to go outside in spite of an overwhelming feeling that a hostile force was watching me. It took every bit of determination I could muster to smoke the cigarette and I only made it halfway through before stubbing it out. I wanted to hide, but from what I didn't know. I opened the door and staggered back into the house. I ruled out driving to the hospital. I knew I would drive too fast or pass out at the wheel. I thought about calling 911, but was afraid the operator would laugh at me.

"What exactly are you afraid of, sir?" the voice would ask.

"Nothing. Please send an ambulance."

"Sir, this line is for emergencies only. What is the nature of your emergency?"

"I'm scared shitless."

"Of what, sir?"

"I told you. Nothing."

I didn't want that conversation to take place. I suppose I could have said I was having a heart attack. For all I know, I was. I didn't make the call. I went to our room and fished out a thiamin tablet. I took it and lay on the bed, bedspread clenched in my fists, closed my eyes against the terror flowing over me, and forced myself to breathe slowly—counting to ten before each release. The fear lasted twenty minutes.

There were other episodes prior to that one. It took me a while to figure out the attacks weren't madness, and not imagined. *Forever.* That's how long the guy at the recuperative care facility said I had to take thiamin. At $4.50 a bottle for a two-month supply, I think I'll keep taking it.

One more thing before I let you go…

I'm not sure the red train is real. It may have been a symbolic conjuration of a damaged brain. The train may not be real, but the destination is. Here the rules of logic fail and faith takes over. I believe there is a Hell. It is a real place, as real as the one you're in right now. But there's good news, too. If there is a Hell, there is a Heaven. I hope I've seen all I'm going to see of Hell, and plan to spend eternity *eventually* in Heaven. Next time I'm going to take the white train—no matter how long I have to wait in line.

Let's not get ahead of ourselves. This isn't about trains. Not yet, not really. Frankly, I'm not sure what it *is* about. I only know I had to write it, and thank you for reading it.

We now return you to your regularly scheduled book. Please remember to keep your arms and legs inside the vehicle at all times…

Section Three – Recovery, Aftermath, Job

I think it's important to tell you this, and it might be hard to believe, but it's important to know:

Within a couple of hours after I got home, I had no *memory at all of any of the events in the previous section.*

Hard to wrap the brain around that idea, isn't it? It is for me. I spent almost two weeks in neural intensive care, and recuperative care: fought enemies that didn't exist, met my wife again for the first time, and all the other things you just finished reading about, and *I didn't remember any of it.* Not then. Memory came later.

§
Red Lobster

I walked in the front door of the house, much like anyone would walk into their house after they've been gone a few days. I looked for my wife and found her in bed. She wasn't feeling well. For the sake of her privacy, I won't explain more than that. She's fine now.

She kissed me and hugged me like I'd been away for a while, as she does when I've been off at a conference or training week. Mom and my stepfather were taking me out to lunch. Sarah didn't feel well enough to go with us, but asked that we bring her something from the restaurant.

We went to Red Lobster. I remember being a little nervous because I knew I couldn't have beer with the meal like I normally would, but didn't want my folks to know I had been ordered not to drink. I didn't know *why* I couldn't drink, or *who gave the order*, but I knew I couldn't drink. I ordered a Coke, or an O'Douls, but I don't remember which one. Neither my Mom or Stepfather commented on it. I don't remember if they ordered alcoholic beverages or not. I'm pretty lucky—I don't care if people drink around me or not. I didn't care then, and I don't care now.

The waitress was great. I asked if they had shrimp scampi. My wife loves shrimp scampi. She said they did, and I ordered some for me and said we would want an order to go for my wife. When she came back with the drinks, I ordered shrimp scampi to go…for my wife. The server took our orders. I don't remember what I ordered, but whatever it was, I asked for another order to go...for my wife.

When we left the restaurant, we had three or four bags—half-eaten meals, and meals to go. My mom and stepfather were terrific. I could tell they were amused (instead of horrified, thank God) by the orders I placed for my wife, who, as you might recall...wasn't there. In fact, my stepfather started suggesting I order things for my wife. It became kind of a game. I'm sure the poor server was tipped well. I'm equally sure she thought I was nuts. Notice the repetition in the previous paragraphs? It's intentional now; it wasn't then.

Sarah and I ate Red Lobster meals from our fridge for the next several days. I can tell you, those biscuits they serve? They're good for about a week, but after forty or fifty of them in a three day period, a feller gets tired of 'em. Even if he can't remember having them a few hours before.

§
Mystery Dots

My balance wasn't great when I slid the door back and got in the shower. It felt good. I hadn't had a shower since before I got sick. I leaned against the tiled wall and let the water run over me. When I washed my arms, I noticed little dots on the insides of each upper arm. On closer examination, I found they were scabs. I'm not allergic to mosquito bites for whatever reason, and I knew I hadn't been outside. I scrubbed. The dots remained.

I'm not sure, but I think they were marks from the probes they stick people with to try to wake them from a coma. I found similar marks between my toes.

David J. Steele

§
Old Habit, No Memory

It was late, that same night. My wife had gone to bed, my folks went back to their hotel, and I was looking for beer. I remembered (incorrectly) that Sarah said she would have beer for me when I got home, but there wasn't any in the fridge. I grabbed the keys and went to the garage to see if it was in her trunk. I thought she forgot to bring it in.

There was no beer in the trunk, or anywhere else in the house. It was Cleveland, and I was sure I could go out and get some. Then I remembered we had some stuff left from the holidays in the cupboard under the wall oven. I was a beer guy, but drank other stuff sometimes. I don't remember what I found under there, but I had a drink or two, maybe three…

When I started to get ready for bed, she was in the hallway. She was beyond mad. She was crying. I felt bad immediately, but I wasn't sure what the problem was. She said we would talk about it in the morning. Still crying, she went to sleep. Still mystified, I did too.

We talked about it the next day. She did most of the talking. I sat and listened, feeling guilty and small, as she told me how close she had come to giving up on us, and how she wasn't going to watch me die again. If I wanted to kill myself drinking, I could and would do it without her around.

"Kill myself?" I asked. "What are you talking about?"

"You almost drank yourself to death!"

Her statement surprised me. I really didn't know what she was talking about. I said as much.

"You just got back from the hospital!"

I was silent. Immediately concerned. What husband wouldn't be concerned if he thought his wife was going nuts?

I wish I was kidding. I'm not. *I didn't remember being in the hospital.* I remembered *she* was in the hospital a couple of weeks before, but I hadn't been in a hospital. I said as much.

She froze. Looked at me the way I imagine you're looking at the page right now.

"You don't remember?" she asked in a very quiet voice. Her tears were gone. Sheer incredulity rode her face. "You don't remember. You really don't remember, do you?"

"No." I felt weak, suddenly. I felt like I was going to faint. We were on the bed next to each other, staring intently at each other's face.

I didn't remember being in a hospital, but I knew she was telling the truth. I hated myself for making her cry, but the fear in her eyes scared me. "What hospital? Why?"

"You almost drank yourself to death. You were there for almost two weeks."

"No, I wasn't. I've been here. I'm on vacation."

"What day do you think it is?"

I wanted to question the question itself. Before I could draw a breath, I realized I didn't know. I had absolutely no idea what day it was, or what I had done that day. After a long moment I said, "I don't know what day it is." The only sound in the room came from the ceiling fan over the bed. It clicked while it whirled.

"I thought you were going to die. You can't drink anymore. The doctors say you can't drink. I say you can't drink. You can't drink. I'll leave you before I watch you die again."

Hard words. She meant them. I could see that. I knew she was right, but I *didn't* know she was right. I couldn't sort my emotions, or the truth right then. "I can't drink." It wasn't a question, but it wasn't a statement either. It was kind of in between the two.

"Never again. It will kill you. They wanted you go to rehab, but I talked them out of it. I'm going to counsel you. We're going to talk about this every day until I'm sure you're not going to drink anymore. If you have to travel someplace, you have to promise me you won't drink. If you can't promise me you won't drink, I'll go with you."

"Can I drink O'Douls?"

She sighed. "Only three a day. The doctor said you can have a couple a day, but no more than three."

§

Doctor Green

I had an appointment to see my doctor. Doctor Green isn't his real name, but I don't remember his name and Doctor Green is in the ballpark, I think.

I was still on vacation. Hard to believe, isn't it? That I went through Wernickes and recuperative care, and didn't miss a day of work? I did. I had been in the Scouting profession for sixteen years, and with that tenure came 21 days of paid vacation.

I remembered the last time I saw Doctor Green. Here's that story…

…*It was before I dreamed I was commanding the Enterprise.*

I was unsteady when I walked into the waiting room. Although I had been to the office several times, I had to stop in the lobby of the multi-storied building and look for a sign, a directory of offices. It surprised me that I couldn't remember where my doctor's office was, but I couldn't. I was wearing an eye patch from Walgreens because of the double vision. I went to the bathroom off the lobby to puke before I took the elevator up to his office.

I wobbled to a chair and waited, biting back bile in my throat. They didn't make me wait long.

Blackout: A Look Inside Wernickes

The nurse took me to an exam room and had me lie on the table. I must've looked like death warmed over. I threw up, but not much came out. It was like a retch with a bonus blob. She got a pan and held it under me. The retches were violent. I was shaking like a leaf.

The doctor came in after only a few minutes. I don't remember much of the conversation, but he asked me a lot about drinking. I lied about it. I *do* remember a few things he said.

"I think it's an inner ear infection, but it could be something much more serious." He was about my age—late thirties—and looked like he might have played football in high school or college. He was usually the kind of doctor that smiles at you when he prescribes antibiotics and tells you to take it easy. Not that day.

"I'm going to prescribe a motion sickness patch, and…" He prescribed something else. Some kind of pill, but I don't remember what it was. "You should feel better by tomorrow. If you *don't* feel better by tomorrow, I want you to call me right away. I'm giving you my home number."

"Why?"

"Because I think it could be something much more serious. Something *very* serious. How much do you drink?"

I lied to him again. He didn't believe me, or I don't think he did, but he was willing to let it go. "I should probably send you to the hospital right now, just to be safe."

"I'll take the pills and try the patch." I maintained eye contact. It wasn't easy. One of the tests he did on me showed my eyes were bouncing from right to left and back again. Nystagmus. It's a symptom of Wernickes. From inside the eyeballs, it looks like double vision.

"You call me or have your wife call me if you don't feel

better by this time tomorrow. Do you have a ride home?" he asked. "I'm not going to let you drive out of here with your vision all screwed up."

"My wife is home. I'll have her pick me up."

"You drove here?"

I didn't tell him I took Carnegie from downtown to his office in the northeast suburbs. I didn't tell him I wasn't sure which lane I was in most of the way—it's a lot of fun driving with double vision. It's a bit of an adventure trying to decide which of the two truck images you see is the real one and which one is the mystery double.

"Yes," I said. "I drove."

"Well, you're not driving home."

"I'll call my wife and have her pick me up." It was a promise I had no intention of keeping. He bought it.

I did try to reach my wife, but she was asleep. I drove home. It was only about ten miles, but it took me over an hour. The reason? ...I couldn't find the place. I circled the neighborhood for a long time, but I couldn't remember where our house was. I filled the prescription at the Walgreens near our house, and then I was able to find my way home.

Like a true addict, I bought beer at the same place I got the prescriptions filled. Went home, stuck the motion sickness patch behind my ear, took the pills with a swig of beer, and turned on the TV to watch while I lay in bed.

Guess what was on television. *Star Trek: Next Generation.*

You know what happened next. The stuff in *Green Goblin* happened next.

Now I was back in Doctor Green's lobby. I wasn't puking. I wasn't wearing an eye patch. I remembered part of being hospitalized, but not much. The events in *Green Goblin* were shifting nightmares and flashbacks of bits and pieces of things

Blackout: A Look Inside Wernickes

I wasn't sure were ever real.

The nurse took me back to the little room. I sat on the edge of the table. Doctor Green was all smiles when he came in. He took one look at me and burst into a grin. "Man," he breathed, "You look great! *Much* better than the last time I saw you!"

I thanked him, but I was a little surprised. It had been a couple of months since my last visit, and most people recover from bronchitis with no problem. I didn't know why I was there, other than my calendar had an appointment with him. I was even more surprised when he started to lecture me about drinking.

"I want you to make another appointment and come back in a couple of weeks. No drinking. You're done drinking."

"What?"

"No bullshit," he said. Visibly forcing himself to relax, he leaned back. "Most people can handle alcohol. They drink every once in a while, and it's no big deal. You're not most people. You're probably one of those people who can't drink at all. It's dangerous for you, maybe even deadly. You're a lucky man, Mr. Steele. Most people who had what you had never recover. Make an appointment on your way out."

I did. I wasn't sure why I needed an appointment, but I made one. On the way home I wondered why, suddenly, people were picking on me about my drinking. It didn't make any sense to me. I felt beat up. Unjustifiably, inexplicably, irritatingly beat up.

§

O'Douls, Sharps, and Buckler

I learned about non-alcoholic beers when I went through the employee assistance program, and after I got home and couldn't drink, I went back to drinking them. Unfortunately, I was still thinking like an addict. I drank more of it than I would have if it was beer. It was an attempt to get a buzz. I hate to admit that, and denied it to myself at the time, but that's what I was doing.

Sarah came back from wherever she was and found several empty cans upside down in the kitchen sink. She hurried back to the bedroom, where I was lying on the bed watching television with a full ashtray and a couple of cans of O'Douls on the nightstand. Believe it or not, I was half looped from the non-alcoholic beer. She was not happy with me.

"You can only have three of these a day." She didn't yell, but there was no mistaking her disappointment. "If you're going to abuse it, you can't have these either."

I was smart enough to know she was right. Besides, the stuff didn't taste good. Buckler came the closest to beer in flavor, but it's pricey…and it doesn't taste right either. I switched to ginger ale. It's safe for me and it tastes good.

§
Email

It was 2005, and I had never heard of Facebook. I don't know if it was around then. There was no access to email to me in the hospital or recuperative care. I asked about it and was told they didn't have access for patients. Makes sense now, and I wasn't disappointed then. I used email heavily, and participated a lot in an on-line Scouting community not affiliated with the Boy Scouts of America in any way other than that its users were volunteers, participants, or people with an interest in Scouting. Like Facebook, I had a lot of regular contacts with people I had never met.

I sat in my den looking at my emails. The list of unread messages was several screens long. I was surprised. I was having a moment, or spell as it were, when I didn't remember being gone. The dates of the messages—stretching back almost two weeks—told me something had happened. I skipped cheerfully over the thought.

The early messages were normal, everyday emails. Then there were a few from my boss. With increasing urgency in his messages, he was asking where I was and why I hadn't responded. Before I got sick, I worked out of my house quite a bit. He was accustomed to having almost instant responses from me. His messages acknowledged I was on vacation and he didn't want to bother me, but wondered why I wasn't responding.

Eventually, he called Sarah. She told him I was in the hospital. Word spread through the Scouting forum I participated in. Many friends from there contacted me after I got out of the hospital, and a couple talked to me while I was in recuperative care.

Looking at the emails, I saw some of what I missed. I missed my high school reunion. I was part of the planning committee, although it was a small part. The reunion was set to occur a few days after I saw those messages. I didn't go. I apologized via email for not attending, and listed coma as my excuse for dropping out of the work when they could have used my help. The others on the committee were, as you might expect, quite understanding.

I called my immediate supervisor—a different man than the one who visited me in recuperative care. I felt a need to apologize, and wanted to let him know I would be in the office on Monday, the end of my vacation, as scheduled.

He answered his phone, and that surprised me a little. Like most good professionals in the Boy Scouts, he was rarely at his desk. I was a little afraid he was going to yell at me. That alone should have told me I still wasn't back to normal.

"Dave!" He shouted my name into the phone like we hadn't seen each other in years.

"Yeah…" I cleared my throat, surprised I was nervous. "Sorry I didn't answer your emails…"

"You sound great!"

I wondered if he was listening to me. I liked the man, but I often wondered if he heard what I was saying. He did, but he didn't always give the impression he did.

"You sound much better than the last time I talked to you. How do you feel?"

"Terrible. I missed the Venturing weekend, and I wasn't checking my email…"

"What? Hey, don't worry about that stuff. You were in the hospital, for God's sake. Joe covered the weekend for you, and it went fine. Thanks for all your work setting that up. I only called because you hadn't been in, and I knew you would be even though you were on vacation, to check registration and all that stuff. When I hadn't seen you and you didn't answer your messages, I called your wife and she told me you were in the hospital."

"I'll be back Monday."

"Don't feel like you have to rush, okay? If you don't feel like coming in, don't worry about it. Just let me know what's going on. We're behind you, man. We're just glad you're okay."

"Thanks." I was near tears. He couldn't tell from the sound of my voice, but I was emotional suddenly.

"If you get a chance," he went on, "call Shirley and Heather. They were really, really worried about you. We were all worried about you, but especially Heather."

Shirley and Heather? It took me a second, but I was able to picture their smiling faces. They were members of our support staff team. Two very nice ladies at the office. I like both of them, but couldn't understand why me being at the hospital for a little while would upset them. "Okay. I'll do that."

I didn't. I chickened out. I wasn't sure I could handle any sympathy. Instead of calling them during office hours, I left them each a voice mail message and said I would see them Monday.

§
Job Search

I spent the next couple of days writing my resume and looking at job sites online trying to decide what kind of job I wanted after I left the profession. One of the messages waiting for me when I got home was from the director of a community center in Midland, Michigan. They were looking for an executive director for a community center in a rural part of the county, and I was one of the finalists. I forgot about the job application I made before I got sick.

I didn't see anything in the Cleveland area I wanted to do. Sarah and I talked and she wanted to go back to Michigan, to Vassar, her hometown. She was sure she could find a teaching job in the area and wanted to be near her mother, who had reached an age where having family close by would be more than a good idea. I agreed with no problem at all. I was ready to leave the profession, and after dragging her around the Midwest for several years, it seemed only fair and right that we do what she wanted.

The job of being an executive director for a small non-profit sounded good to me. I have the background for it. With a couple of phone calls, I had the interview set up. Getting to it was going to be tricky. I had taken my vacation all at once, and I didn't want my bosses, especially after they had been so good to me when I was ill, to know I was looking to leave. I felt like a rat just thinking about it.

§
Doctor Green, again
He broke into a wide grin when he came into the room and found me sitting in the chair. I was happy too. I found his office without having to use the chart in the lobby.

He examined my eyes. I think that's all he examined. Asked how I felt, and if I was still not drinking.

…Not drinking. That's a good way to put it when someone who has done such damage to himself with alcohol quits drinking. Not drinking. For me, during those days and for the next considerable period of time, not drinking was an activity. An activity in the sense that if someone asked me what I was doing, I might have said, "Not drinking."

Drinking was on my mind a lot. Really a lot. Wanting a beer wasn't as bad as wanting a cigarette when you can't smoke, but I did spend a lot of time wishing I could have a beer, or two, or twelve. It's not a fun way to walk around, or sit around…not drinking.

"You're doing very well," Dr. Green was saying. "It's amazing how well you're recovering." Abruptly, he switched gears. His eyes got hard. "No drinking. You can't. If you drink you'll probably relapse, and if you relapse there's a good chance you'll die. Understand?"

"I can't drink." The thought bothered me. "Ever?"

"Not for at least six months. Not a drop." He shrugged. "After six months or so, if you want to have a beer, go ahead and try it. One. Two at the most." He held up a finger. "If you feel funny after a beer, stop drinking right away. If you feel paranoid, get to a hospital. Seriously. You can't ever go back to drinking the way you did before, but you might be able to enjoy a beer. Never more than two."

I've read since then that there is debate among doctors about drinking after Wernickes. Few agree with Dr. Green. Most seem to think that only alcoholics get Wernickes from booze, and that alcoholics can't drink. I agree with the majority who don't agree with Dr. Green.

"You need a cane."

"Why?"

He smiled. "I watched you walk in. Your balance isn't very good."

I hadn't noticed or thought about it before, but he was right. There was a wobble to my walk. It wasn't as bad as it was before I got sick, but it was still there.

"Pick up a cane at the drugstore on your way home. Use it. It's primarily for balance, but keep it with you all the time. I called a physical therapist. He's going to come to your house three times a week for the next couple of weeks, and he'll give you some exercises to do on your own. The more you do them, the faster you'll get your balance back. Okay?"

"Sure." A cane. Surprisingly, I didn't mind the idea. I remembered falling into walls just before I got sick. It wasn't fun.

"Set another appointment on your way out. I want to see you in another two weeks." He shook his head and smiled. "It's amazing. You were in really, *really* bad shape, and now you're back."

§
Staff Meeting

There were changes I hadn't seen yet, and some I had seen but forgot about, when I got to the office. The building was under renovation in the weeks before I got ill, and they were mostly done. There was a secure door to the area where the district executives like me had our cubicles. My key opened the door. I forgot I had a new cubicle by a window. It was a tribute to my tenure and experience. The others were in the middle of the room. Although I appreciated the tribute, it didn't make much difference to me what my cubicle looked like. I spent as much time as I could on the road, seeing people.

Everyone was very nice to me. They greeted me in the slightly uncomfortable way one greets a friend who's been sick —warm, but quietly curious. You want to ask how the person is doing, but you're a little afraid you're going to be the millionth one to ask. I had a cool cane, a folding cane. I tried to act like I really didn't need it, but I didn't take a step without touching it to the ground.

The boss that visited me in recuperative care passed me in the hall and clapped a hand on my shoulder. He had a big, warm, inviting, and genuine grin on his face. "How ya doin', man?" he asked. "You look a lot better than the last time I saw you!"

I blinked at him.

"You don't remember, do you?" He whispered the question. There were others in the hall. "It's okay. They said you might not remember."

I smiled. A piece clicked into place. "That's where the magazines came from!"

"Magazines?"

"When I unpacked my bag, I had a couple of magazines in there. They had your address on them."

"I did. That's right. I brought you a GQ, and a Men's Health."

"Yeah. Thanks."

"Don't stick around long after the meeting," he said. "See Frank before you head out."

I'm glad he said the name. Frank was my immediate supervisor, the one who called Sarah and found out I was sick. The one who emailed me. The one who's name I couldn't remember.

What did we cover at the staff meeting? ...I have no idea. I was there, but I have no idea.

§
Short Long Day

I was tired. Could've fallen asleep in the chair in front of Frank's desk. The afternoon sun was shining on me while I waited for him to finish a conversation in the outer office with someone.

He came in and took a seat behind the desk. "You feel okay?"

"Tired."

"Go home and get some rest as soon as we're done."

I nodded. "What's up?" I was nervous. I remembered he was going to put me on probation if I didn't accomplish some mid-year goals. I couldn't remember if I accomplished them or not. I was pretty sure I had, but only because I always did before. I had learned that assuming such things was a bad idea. "Am I on probation?"

"What?" He shook his head. "That's gone. All gone. We didn't know you were sick..."

"Neither did I."

"Don't worry about that. It's gone. The letter is gone. You were sick. Really sick. That's what I wanted to talk to you about..."

"I'm better now." I jumped in. I didn't want him to finish,

but I knew he would. It was part of what makes us what we are. Professional Scouters don't give up easily on anything.

"You're not there yet. Better, sure. It's going to be a while before you're fully back."

I was glad he knew that. "I have some physical therapy I have to do. They're sending guy to the house."

"Good. Take whatever time you need. Work from home, work from here. I don't care where or when. Just keep me informed and call me once a day so I know you're okay."

"Will do." I wanted to get up and leave, but could see he wasn't done.

"Do you want to go to camp for a week?"

"Camp? Summer camp?" Visions of staying on an old army cot in a cabin, or worse, at camp flashed before my eyes. I used to love to camp, had served ten years or more on camp staff, taught at National Camp School for Cub Scout Day Camps, and administered sixteen Cub Scout Day camps. "Why would I want to do that?"

"It'll be a nice, relaxing way for you to ease your way back. You can stay at a hotel up the road if you want. We're asking everyone to spend a week at camp. It's good PR, and it's good for you guys."

"Okay." I knew damn well he wasn't going to take *no* for an answer, and I was warming to the idea.

"You'll work in the handicraft area. Someone will drive you around in a golf cart so you don't have to try walking the trails with that cane. Get a physical." They were required for all personnel in camp. No big deal…

It was settled. I was going to camp the next week. I wouldn't be camping—unless the Days Inn up the road had a canopy over the bed—but I was going to camp.

The only hurdle came when I stopped at the MedExpress

for a physical. The doctor refused to sign my form. He looked up my records, and when he saw I had been hospitalized for Wernickes, he *refused* to sign the form. Absolutely-no-way-in-hell was I going to camp.

That cinched it. I went home and dug out my physical from the 1997 National Scout Jamboree. That form was signed after a thorough medical examination, and unlike other BSA physical forms...it was good for 10 years... I made my room reservation.

§
Physical Therapy

I didn't recognize the name of the physical therapist on his way to my house, but it turned out he recognized mine. He was the committee chairman of one of the Boy Scout troops I served. A very cool guy. Sarah wasn't home when he came to the house. I didn't want her to be around to see me do whatever it was he was going to make me do. I promised myself I would do whatever was necessary to get rid of the damn cane as soon as I could. I wanted to be normal again. Burned for it. If I was going to be stuck with a cane for the rest of my life, I was going to make sure I did everything in my power to avoid it.

The exercises weren't as bad as I was afraid they would be. Most of them, if not all, centered around me getting my balance back. At first he walked behind me as I did them, ready to catch me if I fell.

We had large eat-in kitchen. I started at one end walking by putting the toes of one foot against the heel of the other foot, then switching my feet, and so forth. Forward and backward. Hopped on one foot across the room, then switched feet and hopped back. The therapist said I had done well, and told me to do the exercises once a day until his next visit. Maybe I'm weird, but I did as instructed. I never did the exercises when Sarah was around, but I did them. I kept the cane with me but tried to use it less and less.

By the time the therapist came around for the second visit —a few days after the first—I was pretty confident in my abilities. At the end of the session, we sat in the living room sipping coffee.

"I'm authorized for five to eight visits," he said. It was a bright, beautiful, late-August day outside. The sun streamed in through the picture window in the living room and a breeze ruffled the curtain behind him. "Today is my last visit. You don't need me anymore."

"I don't? This is only the second one."

He smiled. "Most people don't do the exercises I give them."

I wondered why, but didn't ask. If he knew why they didn't do them, he could probably get them to do their exercises. I saw them as a means to an end, a means to victory. I wasn't going to let Drunk Dave turn me into an invalid. No way, no how. "Those exercises are my way out."

"Keep doing them. Do them twice a day for at least a month, more if you still aren't happy with your balance."

"Can I get rid of the cane?"

He shook his head. "Keep using the cane until your doctor says you don't need it anymore."

I did.

§
Lonely Days

Sarah got a teaching job at the middle school she attended as a kid, and we were both excited. I didn't get the community center job, and I was kind of glad I didn't. I went on the interview still weak from the sickness, hobbling along with my cane. I remembered the territory well, and was amazed at how the community center had grown. When I left the area in 1997, it was a fledgling community center with one building with a big room and a kitchen. A state-of-the-art facility had been constructed on the site. It was beautiful and the area needed it. The interview went very well, until they asked me about sports. I crashed and burned. I'm not a sports fan at all. I don't watch them, and don't care about the outcome.

We had decided to move back to Michigan whether I had a job lined up or not, and once she landed that teaching job, it was a done deal. She brought up again the offer to let me write—long a dream of mine—and take a part-time job. Her salary was enough for us both to live on, although not at the level we lived while I was working for the BSA. We would sell our house (I can imagine your sigh. Selling a house in 2005 was a bad idea, but we didn't know that then), and buy one up there. In a month, we would start our new lives.

It was a hard day...the day I resigned from the Boy Scouts of America. I never thought I would do that.

§
Resignation

I think Frank was surprised when I resigned. It was obvious he didn't want to accept it. Thirty days written notice was required according to our contract. I told him I was giving him thirty days, but I hoped—and expected—it would be a shorter period. In my experience with resignation, even by tenured professionals, we wanted them to leave sooner rather than later. Usually, by the time someone resigns from the BSA, they either have another job or there's a mutual sense of relief that they're leaving. That's probably common to most companies.

He took it well. "I'll have to tell Mike," he said after a pause. "Stick around while I do."

I really didn't want to stick around. I had been dreading the conversation with Frank, and I sure as hell didn't want to talk to Mike about it. They were good men, and I was a good man. My resignation had nothing to do with them, but I was sure by then it was the right thing to do.

He came back about ten minutes later and said Mike wanted to see me in his office.

The Scout Executive of the Greater Cleveland Council has a really nice office. It's in the corner of the building—a building built to resemble independence Hall in Philadelphia—and has windows on two sides. He had a big desk in front of a set of windows, facing the door. A leather couch was against the wall to the right of the door, and a round table with a few chairs around it was by the windows in front of the desk.

We didn't sit. He got up from behind his desk and came to shake my hand. I wasn't sure what he was going to say.

"You're leaving to write books, eh?" His smile was sincere and warm. "I think that's great. I knew a guy who did that, and he was published quickly. We're going to miss you. I wish you wouldn't quit, but I understand."

"I went through a lot, Mike. It's time for me to make some changes, and this is one of them."

He nodded. "I understand."

"I'll give you my best for thirty days."

There was hesitation on his part, and I was sure he was going to tell me that wouldn't be necessary…

…But that's not what he said. He had me talk to Frank again.

To my complete surprise, Frank asked me to stick around

until the end of October. *October!* In seventeen years of professional service, I never knew anyone who was asked to stay around after they resigned for thirty days, let alone two and a half months.

"Mike wanted to let you go right now," Frank was saying, "but I'm asking you to stay a little longer..."

"I'm not going to change my mind," I interrupted.

"We know. We need your help. *I* need your help. You're the best I've got, and I'm already running a vacancy. There's a hiring freeze and the fall is our busiest time of the year. We *need* you to get through the fall."

I told him I would. We agreed that my last day would be October 27th, 2005. The date was significant, but Frank didn't know that. October 27, 1988 was my first day in the profession, and seventeen years later that day of that month would be my last.

"...Go home and write your resignation letter," Frank was saying. "I'm not going to turn it in right away, but go ahead and write it." His eyes hardened in the way of a man who has half a second, suspicious thought, "Don't take me for a ride. I know you won't, but don't. Don't give up on me, because if you do, you'll be gone before you know it..."

I didn't bother to point out he was threatening to get rid of a man who was trying to quit in the first place... "I'm not going to burn you."

"I know. I guess I just had to say it."

"The hell you did."

"Sorry. ...End of October, okay?"

"Yeah."

I cried on the way home. Some would say men don't do that, but this one did.

In retrospect, I shouldn't have stuck around. I got through

it, and wasn't useless, but I wasn't working at my best. I forgot things, sometimes things that bit me in the ass. I stayed out of the office as much as I could—knowing I was leaving made me more relaxed than everyone else in that intense time of the year, and Frank didn't want others to know I was quitting—and went in only long enough to gather materials and turn in applications and money for new Scouts and adult leaders.

They had a going away party for me, a surprise going away party. I can laugh about it now, but didn't give it much thought then… I wasn't there. I *would have* been there, if anyone had told me about it. I didn't go in often, and no one knew when I was going to be at the office. It was sweet of them to have a going away lunch for me, but I'm not in the habit of going to appointments I don't know exist.

I found out about it after one of my brief visits to the office to turn stuff in. I was almost to my car when one of my fellow professionals pulled up in hers. She rolled down her window and said, "We had a party for you…" She looked away and hesitated. "I apologize for not telling you about it. We wanted to surprise you, and didn't think to let you know about it." Before I could say anything, she handed me a gift-wrapped box. "This is from us. We'd like you to keep it next to your computer when you write."

I thanked her and asked her to thank the others, for the gift and for the party. I appreciated the thought, I really did. Opened the gift when I got home. It was, and is, a pewter little train set. A nick-knack, a BSA specialty item. I like it.

§
Doctor Green for the Last Time

This time when I went to see Dr. Green, I was wearing a suit. I was on my way to the office. He broke into a wide grin when he came into the room and found me sitting in the chair. I was happy too. I found his office without having to use the chart in the lobby.

He examined my eyes. I think that's all he examined. Asked how I felt, and if I was still not drinking.

"You're still not drinking." It wasn't a question.

"That's correct." I didn't tell him not drinking sucked.

"You're doing very well," He was grinning and looking at me like he was seeing a miracle. "It's amazing how well you're recovering."

"Thanks."

"Are you still doing your exercises?"

"Yes."

"I'm setting an appointment for you to see a neurologist, and a neuropsychologist. They'll help you with the rest of your recovery."

"We're moving to Michigan."

That threw him a curve. "When?"

"She's already up there. My last day is in a couple of weeks, and I'll be up there too."

"Where in Michigan?"

"A small town near Saginaw."

He asked me for the names of hospitals in the area, and left the room when I told him. When he came back, I could see he wasn't happy. "We don't have a network up there, so I can't give you a referral. Go see a neurologist when you get there. If you need a referral, just call me and I'll give you one."

I would have done that…if I remembered. Once I did remember, I couldn't remember *why* I needed to see a neurologist. I was fine. Correction: I *thought* I was fine. It would be months before I figured out I wasn't. And months more before I found someone who could help.

David J. Steele

§
Initial Healing, New Phase of My Life

It was dark when my wife kissed me goodbye. I didn't know where I was, and I didn't know where she was going. I asked both questions.

"School!" she said. I could hear her smile even though I couldn't see it. She was lit from behind with light from the hallway, and the bedroom was dark. "We're at my Mom's. We live here now, remember?"

It took me a second, but I remembered. We were living there for a while until our house in Cleveland sold. I wasn't a professional Scouter anymore. I was a husband, living with his wife at her mother's house. I was going to get up soon and go downstairs to write for the day.

"See you tonight," I said. "I love you."

"I love you, too."

§
Learning

If there was a better environment for writing, I had never seen or heard of it. Mom, and by that I mean Sarah's mom, made me a place in my late father-in-law's home office. The house is a big house, a really big house, built on a hill overlooking woods, with glimpses of a river available through the trees. There's an indoor pool, a big kitchen on the ground floor above that overlooks the pool, and a large living room on the other side of the dining room above...all overlook the pool, and all rooms have a view of the woods through huge windows or the patio doors off the pool.

The office in the house was full of books in cases along the parts of the walls that didn't have windows. There were windows in front of the desk, and windows to the right and left of the desk. My computer was under the desk, and on the desk was my keyboard and monitor.

There was only one problem, and it was a big one. I couldn't write. I didn't want to admit that to anyone, especially Sarah. During the sixty days before my resignation took effect, I tried to write fiction, but I couldn't do it. Always before, images came to my mind in ordered fashion and I only had to record what I saw with my mind's eye. Dialog always came almost on its own.

My mother, one of my step-sisters, and my brother came to the house to pack before we moved. I was told to stay out of the way, and was only in charge of packing my home office in Cleveland. They packed just about everything in the house in two days, but I couldn't manage to pack the contents of that den. My mind was too disordered. I tried to write then, and couldn't. I wrote thank you notes for all of them, and that minor task took me hours *for each note.*

And now, for the third or fourth day in a row, I went down the stairs at my mother-in-law's house to the office space where I was going to write my first bestseller and fired up the computer. I looked out at the woods, full of green life. Through the windows to my left, I could see bird feeders on wooden posts. Wild turkeys live in that part of Michigan, and they have a loose routine. I knew they would come sometime between nine and ten in the morning and peck at the food under the feeders. They're fascinating to watch, those big wild birds. Usually, three or four of them would come. One or two would serve as a look-out while the others ate. They're big, but they're far from dumb. One walked up to one of the windows once and stared through the glass at me. The intelligence in its eyes was eery.

I tried to write a chapter. Chapter one of a book I was

David J. Steele

going to call *Sexton*, about three American teenagers who find a gateway between worlds and go through it. I typed, "Chapter One," and stared at the blinking cursor under the words.

Stared. Nothing.

Nothing.

I tried to do some free writing—just typing whatever came to mind—to get the old juices flowing. What came out wasn't gibberish, but it wasn't any good either. I stopped, went upstairs, poured a cup of coffee, and went outside to smoke a cigarette and try to think of anything but my inability to write.

When I finished smoking, I went into our bedroom and dug out a paperback I re-discovered before I left Cleveland. I'd read it before, but hadn't looked at it or thought of it for almost twenty years. It was called *Magician Apprentice*, and it was by Raymond E. Feist. It's part of a series, and it's excellent. Inspiration of a sort struck: I couldn't write, but I could type.

Maybe, *maybe...*

In 1986, as a Reagan Scholar at Eureka College, I was given a unique opportunity. Even then I dreamed of being a writer. The Reagan Scholarship program was a great boon to me as a young man. It was a full-tuition scholarship based on leadership to Ronald Reagan's alma mater, Eureka College in central Illinois. Reagan Scholars were given mentorship opportunities—usually intense job-shadowing experiences—with notable people in the field of the scholar's choice.

I wrote a lot in college. I had a column called *Machiavelli's Corner* in the weekly campus paper. I wrote short stories in my spare time. I had been working on a novel and for whatever reason grew frustrated with it and threw tossed it in a wastebasket in the student union. The director of the Reagan Scholarship fished it out of the wastebasket without my knowledge and liked what he saw.

He suggested I might want to meet some real authors and learn from them. I liked the idea and gave him the names of my favorite authors at the time. Reagan was in office, and the scholarship and its director were able to set up a lot of really terrific opportunities for us. I gave him several names, but at the top of the list were: Clive Cussler (most famous at the time for *Raise the Titanic!*), Frank Herbert (*Dune*, and others in the series), Terry Brooks (*Sword of Shannara*, and several others), and Stephen R. Donaldson (*The Chronicles of Thomas Covenant the Unbeliever*, and others.)

The only one I didn't meet was Frank Herbert. He had an excellent excuse for canceling: he died. The scholarship purchased an Amtrak All Aboard America pass for me, and for three weeks that summer I could ride the train anywhere in the country. When you're twenty, as I was at the time, you don't mind traveling by coach. It was an invaluable experience.

Clive Cussler was awesome. He let me stay with him and his wife in their beautiful home near Denver. For a couple of days, I wrote. The routine was simple, but highly effective: I wrote a few pages, took them to him for critique, and went back to the guest room to revise. He was gentle but firm, brutally honest but appreciative of my ability. We kept in touch by letter for several years after that.

Now, twenty or so years later, while trying to recover from a serious illness, I couldn't string words together and have them make sense two hours after stringing them.

...But I could type. I could type fast. A hundred twenty words a minute was about average for me. It's better than that now.

I took my copy of Feist's book down to that office overlooking the river, folded back the paper cover farther than the printer intended it to be folded, and stuck it on a typing stand. I started to type what I saw. Everything: title page, copyright, dedication. Word. For. Word.

Clive told me about a guy who wanted to learn to write, a man he knew from a trip on a ship. The man copied Herman Melville's *Moby Dick* by hand. When he was done, he tossed the pages overboard and claimed he knew how to write. When Clive told me about that exercise, I laughed. We joked that the man only knew how to write like Melville.

I'll tell you, though…when you can't think of anything else to do to spur what used to come so easily, copying someone else's book seems like a pretty good idea.

I typed for weeks. I read. I typed all of Feist's first three books. I typed books I found in my father-in-law's office. I typed travel brochures. I typed articles about writing. I typed Ambrose Bierce's *Devil's Dictionary*. Eight or more hours a day, seven days a week for a couple of months, I typed. I produced no works of my own.

Sarah knew, perhaps better than I did, what I was doing. She was very understanding, considering I left behind a solid income to write and had yet to turn out a single page of my own. Her patience had a limit, though, and I was close to reaching it. One night, as we enjoyed a private moment before turning out the lights, I confessed I didn't know how to write anymore.

"Don't tell me that," she said. "You have to write. You have to!"

§
Dreams

I have read a lot about Wernicke-Korsakoffs. I've learned a lot about Wernickes, but nothing I have read says anything about dreams. You can find a lot about dreams on the internet, and most of it relates to dream interpretation. Most of it is hokum, as far as I'm concerned.

I can tell you this: I didn't dream much when I was drinking. I dreamed vividly, with background music, and color, and sometimes sensation, before I started drinking. Dreams plagued me when I was in the hospital and during my time in recuperative care. I thought it was due to the drugs they had me on, and it might have been, but I wasn't on any drugs while we lived at my mother-in-law's house.

The more I typed, the more I dreamed. I think I was awakening parts of my brain that had been dozing for years. I think all those words that translated into images while my fingers flew over keys—keys that were losing the letters on them because I was typing so much—triggered the dreams. I felt better when I woke up. I could remember bits and pieces of my dreams.

Aren't dreams the product of a brain working in slumber? I think so. Research seems to indicate that dreams are an important part of our mental health, and I was finally having them again. It's still true for me: the more I write, the more I dream, and the more I dream, the better I can write.

§

The Godfather—Breakthrough

...by Mario Puzo, I typed.

It was gentle pressure, the pressure I felt from Sarah, but it was pressure. I knew when I folded back the cover on that paperback copy of *The Godfather,* that the book wouldn't survive me sticking it on that typing stand until I was finished with it. It was twenty or more years old and had never been read. The glue holding the pages together was dry. It crackled when I looked inside.

I typed, and typed, and typed…then something happened.

At some point in the novel, I started to change sentences. Here and there, sporadically, I made changes. Subtle at first. I played with bits of dialog. I rearranged paragraphs, and trimmed description in some areas while adding it in others. I felt guilty about it, as if I was somehow betraying the author by playing with his prize-winning book.

If you're thinking you would like to see a copy of what I did to the book, forget it. When I was done, I deleted all copies.

I was writing again. I was ready to start my own stuff.

§

Writing *Green Goblin*

Some of my dreams seemed like flashbacks to my hospitalization. They terrified me. I was epileptic as a kid, and I was born with my right foot angled away from my knee. As a little kid, I had my fill of hospitals and doctors. My baby shoes have holes in the soles where a bar screwed in. The bar was there to "train" my right foot to go straight out from the leg. It must've worked. I grew out of the epilepsy, by the way. I had three grand mal seizures. My epilepsy wasn't inherited. It came from scar tissue I got on my brain as a little kid when I tossed a rock in the air and caught it with the top of my head. Even then I was lucky when it came to health. When my head grew, the scar tissue got thinner and thinner and eventually became a non-issue.

One night as I stood outside in the dark and smoked my last cigarette of the day, I stared at the tip of the cigarette. In the orange-red cherry of the cigarette, I saw the foothills of Hell again.

I had to write it. I had to record as much of my experience as I could. I had to do it for me, and I had to do it for people who needed to know more about Wernickes. The *idea* of writing it scared me; I won't kid you about that. To write it, I had to re-live part of it. I didn't remember much then, but I knew it was scary stuff. I did it, as you know. I wrote *Green Goblin* and shared it with Sarah. I shared it with *Granta*, the leading literary magazine of the world. They rejected it with a personal letter I found quite flattering. They praised the writing and me for writing it, but said it's just not their kind of story. I was honored.

§

Emotions

Alcohol has an effect on emotions. You don't have to be a rocket scientist to know that. I usually had a good grip on my emotions, and when I was sober, I don't know anyone who would have said I didn't. So I was surprised at the emotions that gripped me so often, and so tightly, in those months after getting out of the hospital.

I was a middle manager, and was used to feeling frustration from those I supervised as well as from the one who supervised me. I was used to taking it on the chin for stuff I didn't do or did do, or stuff other people did or didn't do. I could accept all of that. I didn't like it much, but I could accept it.

As a result, I was surprised on more than one occasion when I got angry, or felt fear, or was so frustrated with something simple that I knew I wasn't myself. I couldn't handle it when I sniffed anger or frustration from Sarah or her mother, whether it was real or something I imagined. On more than one occasion, I thought about running away—that they would all be better off without me if I just walked to the end of the driveway and kept going until morning, and morning, and morning.

When Mom and Sarah went to bed, I went online. I had a feeling my emotions had something to do with Wernickes Encephalopathy, which by then I knew I had gone through. I'm good at online research. I don't give up easily, and I can read and get something from any article, whether it's written in medical-speak or plain English.

Thiamin. The absence of it leads to Wernickes, and Wernickes almost always leads to Korsakoffs. It didn't take a lot of research to discover I had been lucky, or blessed. Probably both. I had a vague recollection of getting instruction about vitamins I had to take after I got out of recuperative care. I went to the storage unit (one of three, actually) that had boxes full of our stuff. It took a couple of days of digging, climbing over stacks of furniture and boxes, but I found the folder with some of my records in it. The written instructions called for me to take a multivitamin every day. I was doing that, but those don't have much thiamin in them.

I went to the local drugstore and found a bottle of thiamin. I started taking it when I felt afraid or angry, or any other strong—and by that I mean *overwhelming*—emotion. It helped. I wasn't getting much thiamin, if any, as part of my diet. My mother-in-law ate mostly fish, and chicken. There isn't much thiamin in those. Beef has a lot of it, and I knew I felt better on those rare occasions when we had steak or burgers.

I applied for a lot of part-time jobs. The economy was going down the tubes, and it was a bad time to apply for work as well as a bad time to sell a house. Our house in Cleveland hadn't sold. The realtor was pressuring us to lower the price again. We bought the place with $35,000 down and every time we lowered the price we were carving into our equity. On the other hand, the house was draining us. Mortgage payments, electric bills, water bills, paying the kid across the street to mow the lawn, were all having a negative effect on our cash flow.

Sarah pressured me—and I'm glad now that she did—to get a part-time job. It wasn't as easy as I thought it would be. Part of the problem was that I had a lot of management experience and a college degree. Those aren't assets when you want to work thirty hours a week for someone else. I applied for a lot of things: executive director for the historical society in Frankenmuth, security guard for Sears, 911 operator for the county. I think I came close several times, but hadn't been hired.

I didn't feel as much pressure as I should have until Sarah told me she wanted me to find work not so much to help pay the bills, though that was certainly part of it, but because she thought it would be an important part of my healing process.

She was right about that. I had reached a plateau.

I find "plateau" to be a scary word. It's a flat-line word. I started writing *Sexton*. Sarah liked it. I liked it. I self-published it eventually, and it has received 5 stars out of 5 stars, so far, from every reviewer.

Sarah told me to apply at the Bavarian Inn Restaurant in Frankenmuth, Michigan. It's a big restaurant famous for its fried chicken. I knew a little bit about restaurant work. I was a dishwasher at a Bill Knapp's Restaurant for six months when I was in high school, and I knew working in a restaurant is hard work. It doesn't matter what job you have in a restaurant. They're all hard work.

I applied. I minimized my experience on the application, and I don't think I admitted having a bachelor's degree. The human resources manager came out to greet me in person when I handed in the application. I had checked the box for housekeeping as my first choice, and I think that surprised her a great deal. She said with my experience, I should be in the front of the house. "Front of the House" means the service part of the restaurant—the part that deals with guests. That was the *last* thing I wanted. I didn't trust myself to maintain a pleasant exterior under pressure.

She called me a couple of weeks after I applied, and I went in for an interview. I was hired as a housekeeper—as you know from reading the Green Goblin portion of this book. I think taking that job at that place at that time was one of the best things I could have done…

…For the first time since getting sick, I started to feel good about myself. Don't get me wrong: that part wasn't there on that first day when I held a johnny mop in my hands and swiped my nineteenth toilet of the day, the day that started when I punched in at 3:57 AM. It's a hard journey to go from Reagan Scholar, to hot-shot up-and-coming executive, to manager with a bright future, to hospital, to janitor. It was a journey that I didn't know at the time was saving my life, but it's a hard journey. It's every bit as fun as it sounds, but I was able to make it fun most of the time and found satisfaction in doing a good job.

Section Four—Korsakoffs: Long Road to Diagnosis

Our house in Cleveland finally sold, and we got a great deal on a big house about six miles from Sarah's mom. It was close enough that we could be there in less than ten minutes if the need arose, and far enough for us to have our own lives. It's a good arrangement. We lost every penny of equity we built in the previous four homes we owned, but we got a great price on a house with five bedrooms, a den, formal dining room, and huge living room.

I was doing well at the housekeeping job. My employers were pleased with my work, and I had never taken a sick day. When four o'clock in the morning rolled around, I was there and ready to work. It was hard work. It was dirty work.

It would have been *smelly* work: cleaning old chicken grease off the floor, mopping the trash compactor room, and other less pleasant duties...but by then I noticed I couldn't smell. That sense was dead. I didn't mind, under the circumstances. Sometimes now it flickers into existence, but then it fades away. I'm okay with that, by the way. It gives me an excuse to leave the litter box alone in spite of the protestations of the cats and my wife.

I was starting to notice problems with my memory, but I didn't realize they were problems with my memory. When someone forgets something, a reminder can be a big help. We've all been there and when we're reminded, a little light bulb goes on and we say, "Oh, yeah. I remember." That's not always the case with me. There are a lot of times when reminding me of something doesn't work. The memory is gone. That's a Korsakoffs thing. Memory impairment.

There was a list of jobs to be done every day in housekeeping. If it was yours to do, your initials were next to the task on the list. I did them all. I got very frustrated when anyone changed the list—which some of the housekeepers liked to do, especially if it meant they got to stick someone else with the nastiest of tasks. I didn't mind the nasty tasks—especially since I couldn't smell. I liked the nasty tasks because I could do them alone. What I didn't like was the surprise, the disruption of my planned order of nasty tasks.

David J. Steele

I didn't have a regular doctor. When I got injured on the job…by bashing the hell out of my elbow while scrubbing something, I went to the local MedExpress. The doctor siphoned 50 mg of fluid out of my elbow. We have medical insurance through my wife's employer, so it wasn't a workman's compensation issue for me. For the next two weeks, I wrung mops out one-handed, but I didn't miss any work. I was injured another time on the job when I was too busy looking at a gap as I stepped between two balconies while washing windows to notice the overhanging gutter. Bashed my forehead when I bent down, and bashed the back of my head when I fell backward. Knocked myself out cold.

Forget seeing little birds when you do that, by the way. I think I had time to spit half a cuss word before the lights went out.

I started to experience pain sometimes. Pain in my legs, as if I had lifted too much. I blamed it often on lifting things, but could never remember any specific start to the pain. I felt the pain in my balls, too. I went to the MedExpress several times for pain. They prescribed fentanyl (which is pretty strong stuff), and I was afraid when I saw the warning about its addictive properties. I was sober and didn't want that to change. I cut the pills in half, and flushed some of them before the prescription ran out. The MedExpress doctors strongly suggested I find a general physician to see. I tried to find one in Frankenmuth, but there was a waiting list.

One day the testicle pain was bad enough that I left early and went to the MedExpress. The doctor there inspected my testes, and ordered an ultrasound. It was an immediate, you're-going-to-the-hospital-for-testing-right-now situation. They suspected testicular cancer. When it wasn't that, they suspected a hernia. They wouldn't let me do housekeeping work.

My employer let me work in the office, and it didn't take long for them to realize I had a brain. After a week of working there, and no discovered hernia or cancer, I was cleared to go back to work in housekeeping. By then an offer to work in the accounting department of the restaurant was made and accepted. I moved to the department that controls the cash for the various tills, and started working nights doing a variety of data entry tasks, giving out change and starting cash, and other related stuff.

I was still in pain, particularly when the weather changed. I had given up on thiamin, and thought it might have something to do with B12 anemia. I was wrong about that, but there wasn't anyone to contradict me. I started taking boatloads of B12.

The next time my balls hurt, the MedExpress referred me to a urologist. He didn't find anything wrong, but suggested a very good doctor in Reese, MI, not far from my home. *That guy—and he's still my doctor—is great!* He ordered blood tests, and sent me to a neurologist.

§
The First Neurologist

I was negatively impressed when I sat in the lobby well beyond the time of the appointment. It generally pisses me off when people are late for appointments, and I'm willing to cut doctors a little slack because they don't always know how long an appointment is going to take, and time is money, so they stack them close together. The leeway I'm willing to give them is about ten minutes.

He finally took me in and inspected me. My main complaint at the time was numbness and tingling in my toes on my right foot, and cold fingers when they shouldn't have been cold. He tapped my toes and fingers with a tuning fork and asked me to tell him when I could no longer feel the vibrations. When he was done, he said, "There might be a little neuropathy in that foot, but not in your fingers."

"I had Wernickes Encephalopathy," I said. "Is it related to that?"

He looked me in the eye and was wrong, dead wrong, with what he said next:

"Forget about Wernickes. You're *over* it."

I asked about thiamin.

"You get enough from your diet...unless you're drinking, Are you drinking?"

I wasn't. I told him I wasn't. The question wasn't insulting because I was expecting it, and I was answering truthfully.

"Thiamin won't help you. You get all you need from food." He shrugged. "You can take it if you *think* it helps. It won't hurt you, but it won't help you either."

I'll sum up what I think of his assessment in one ugly word: *Bullshit.*

∫
EMG—ouch!

My doctor was as unhappy with that neurologist as I was, but not because he suspected Korsakoffs. He ordered an EMG, and his tone when he told me he was ordering it was vaguely apologetic. I'd never heard of the test. I asked him what it was. I had EEGs when I was a kid, and those didn't bother me at all. That test scans your brain's response to electrical stimulus with little probes. It's no big deal. I thought an EMG would be similar.

Not so. He said the test involved some pain. I researched it on the internet, and every site said it was a painful test. I've been in pain, and it passes. I wasn't very worried about it. I was worried enough about it to look up what an EMG is on the internet, but that's not saying much. My job at the time involved sitting in a little room waiting for the last money for the restaurant to come in at night. While I waited, I researched things on the internet.

An EMG involves testing the nerves by sticking needles in them and then zapping the nerves with little bits of electricity. Every article I read about EMG tests say it's a painful test. One said it was like getting a tattoo. I don't have any tattoos, so I don't know what one feels like. I shrugged it off. Lots of people have tattoos. Some people have lots of tattoos. Besides, there was no prep to worry about.

I knew about prepping for those tests. My doctor had ordered a cat scan of my legs to see if there was any reason for the pain. I'm glad I took that test. It revealed small partial blockages of my arteries in my lower left, and upper right legs. If it wasn't for neuropathy, we wouldn't know about those. We keep an eye on them with Doppler tests. Doppler tests are fun. I get the giggles when they squeeze the arteries in my legs and I can hear them *sqreeeeek* shut. No pain with a Doppler test.

The EMG sounded like a different kind of thing. Needles, electricity, and me. Nonetheless, I arrived at the appointed hour. On my way in, I asked the young lady escorting me if the test was painful. I gave my example of tattoos. She had one.

She said, "Oh, this thing hurts a *lot* more than a tattoo." I wondered if messing with people was part of her job description, but decided it was probably just a nice perk.

I sat on the patient table in the room and waited. There was a device on a table next to the bed that looked like some sort of WWII radio built into a white case. There were wires coming off it. Big dials with encrypted figures on them, little chart thingies, etc. I tried not to study it. I knew if I looked at it too long, I would try to tune in Radio Free Europe.

It seemed I was the only one, at that point, who thought my leg pain had something to do with Wernickes, but they were doctors, and I wasn't.

The technician came in, wearing the usual white jacket. He looked to be of oriental descent and did a good job chatting with me while sticking little needles in the nerves of my arms and legs, and…*ooohooooooch*…the back of my neck. Turns out he was Korean, and something of a philosopher. He asked philosophic questions, and I enjoyed the conversation. When he turned on the juice I found out all the stuff I read about EMGs was true. They hurt! It wasn't excruciating pain, but it's pain enough that if someone asks you if you want to be hit upside the head with a ping pong paddle or have those tiny little needles stuck in your arms and legs, go for the paddle. It'll hurt more, but you'll be done before you know it.

I was sitting there with needles stuck in my arms and legs and the back of my neck, juice running down the nerves in my right hand…and his cell phone rang.

He looked startled and pulled the device out of his pocket. He stood up and said, "It's my daughter. I'm sorry, but I have to take this."

He left the room. He left the electricity on. It burned a little, then it burned a lot. Then I felt like my fingers were on fire. Then it hurt. Then it hurt a lot. I started to feel steam coming out of my ears. I bided my time enjoying images of ripping the needles out of my arm, wrapping the wires around his neck, and sticking the needles in his ears while trying again to find Radio Free Europe on the EMG machine.

It was seven minutes before he came back. I looked at my

watch and waited. I had some choice things to say to him, but the look on his face made it unnecessary. He was shaking his head at his daughter when he opened the door, but one look at my face had him gasp. He ran over to the machine and started turning it down in a flurry of fingers.

I asked if I had neuropathy. He said, "No. You have *poly*neuropathy. It's all through your arms and legs."

I discussed the test later with my doctor...the general practitioner. The one who *doesn't* bring his cell phone to the exam rooms. We both had a chuckle about the guy leaving the juice on when he took a call, but I'm pretty sure my doctor reported it to the hospital.

My doctor ordered an EEG—a test I had as a kid after my grand mal seizures, and again when I was older. The EEG results were clean. I'm not sure what they were looking for, but my results were normal.

§
Leg Pain, Research, and Relief

I researched Wernickes and Korsakoffs in greater depth than I had before. I saw an article someplace that said the neuropathy from Wernickes could be treated with vitamins, and it made sense to me. I was taking thiamin every once in a while to fight the paranoia feelings I had from time to time, and it seemed to help. It was a vitamin, and I knew firsthand what happened if the brain doesn't have enough, and I knew there's no such thing as *too* much, so I took one tablet in the morning and one in the evening.

I was fortunate enough to see a post in an online support group by a person who works for a place in the United Kingdom that treats people with Wernickes. I'm happy to call her my friend. She told me that some patients respond very well, after they leave the hospital, to large doses of thiamin. She told me they take 1,000 mg a day or more, and vitamin C to aid in the absorption.

I started taking more thiamin. A lot more than one tablet in the morning and one at night. Now I take four 100 mg tablets when I get up in the morning, three 100 mg tablets after I've been up for a couple of hours, three 100 mg tablets in the mid-afternoon, and hope it gets me through until bedtime. It works for me. Most of the time, unless there are changes in the weather, I'm pain free. I don't take thiamin late at night because if I do—and I haven't found any medical research beyond my experience with this—I have dreams that are so vivid they wake me up.

It makes sense to me on this level: lack of thiamin got me into trouble, and having thiamin in large doses keeps me pain free. I don't need medical insurance to cover my thiamin purchases—and that's good because I don't think it will. A bottle with 100 thiamin tablets costs me $3.99 at the local drugstore, and only $4.99 at the grocery store. It's not a placebo either, no matter what that first neurologist said. I tried to using B12 and B-complex vitamins. Pain returned in 48 hours.

Ultimately it wasn't neuropathy that led to a diagnosis of Korsakoffs. It was memory, or lack thereof.

§

Never Thought I'd Be Glad to Get A Bad Performance Review

I knew I was having problems not forgetting things (slightly different than remembering), but I didn't know how visible or how much trouble they were to others. There was a project, an accounting project, my boss had me work on. It was going to take several hours to complete and span three or four of my shifts. It's a part-time job, and I had a couple of days off after I started the project.

It's a shared work space, and when I returned after my days off, there was pile of papers in my way. They didn't look familiar. I muttered about slobs in the work room and put them out of my way on a shelf.

Several days went by, and one day my boss came into the room as I was about to leave. She said I couldn't go until I finished the project. She was somewhere between irritated and angry. She opened a spreadsheet and put the papers next to the monitor. "You're almost done," she said. "Finish it up and then you can go home."

I looked at the papers, sure I had never seen them before. I asked what I was supposed to do.

"Very funny," she said as she walked away. "You're the only one working on it."

I thought that was a strange thing to say. I went to a computer and tried to figure out what to do. It looked like I was just entering a bunch of numbers from one spreadsheet to another. As I did, Excel caught a couple of math errors. I fixed them. It took me about half an hour, and I found a few anomalies.

The weird part—I had no recollection of that project, but I knew what to do. She gave me a funny look when I took the completed work to her and said I wasn't sure I did it right, but after I checked it against the work of whoever started the project, I thought I did it right.

A couple of weeks went by, then when I was on my way out, She gave me a review, and someone else from our little team sat in on it. My performance was excellent in most areas, but they wanted to talk to me about my memory. I would have received an excellent review, they said (and thankfully put in the review) if I didn't forget so much.

I was upset at the end of the review, but not with them. I had been forgetting things without knowing I was forgetting things. When I got home, I called my doctor and made an appointment for the next day. Then I went to the library and researched memory problems. At that time I thought I was over Wernickes so I didn't bother to research that. I thought something else must be going on.

One of the things I learned is that doctors, at least in the United States, typically don't do anything when a patient complains about memory problems. They rely on educators or bosses or other doctors or their own observations, to report memory problems. That sounds bad on the surface, but it makes sense when you consider that someone with memory problems will probably be an unreliable witness to their own problem.

I had a copy of my evaluation with me when I saw my doctor. His reaction scared me a little. He tried to hide his concern, but he was worried. Really worried. He referred me to another neurologist. It was a week or two before I could get in to see him, and the time between visits was unpleasant.

It was unpleasant because I couldn't trust my memory. I didn't know what was going on, but I'll tell you it's scary when you know you can't remember things you should remember, but can't tell what you forgot. It started a series of self doubts. I questioned myself a lot, wondering what I was going to forget or had already banished from my mind by accident. I worried it would get worse. I wondered if I was going to be disabled after all, or die, or one then the other. I prayed a lot.

§
Second Neurologist, and Three Words

I was getting tired of doctors, and of medical attention. Who wouldn't be? It seemed like something was going on with my head, and no one could tell me what it was.

It was raining when I got to the part of the hospital that contained the neurologist's office. My neuropathy is worse when it rains than it is at other times. Only recently did I learn that humidity plays some sort of part. I don't understand why it does, but it does.

I wasn't sure what this guy was going to do, but I figured hitting me in the hands and legs with a tuning fork wasn't so bad…and if that's what he wanted to do, I was fine with it. If he suggested an EMG, I told myself, I'd let him give me one on one condition. I would tell him he had to take one at the same time I did, and he could control my dials if I could control his. I'm kidding, for the most part.

He came into the room and shook my hand. I wish I could remember his name so I could recommend him to others. I don't even remember what he looked like.

"Did you drive yourself?" he asked. "Find the place okay?"

"No problem. I used MapQuest."

We chatted a little about my health history. I mentioned Wernickes, but he didn't say anything about that. He said he was going to administer a couple of tests by asking me questions. I don't remember the first two, but I remember the one that involved three words. I'd seen it done on Law & Order.

He said, "I want you to remember three things: acorn, firetruck, and picnic table."

"Okay."

We chatted for half an hour or so. I don't remember what

we talked about. I knew he would ask me what the three words were, so I played them over and over in my head like a mantra. I think he could see or sense that's what I was doing—maybe most people do what I was doing. He was a good conversationalist, and I found myself enjoying whatever it was we were talking about.

"What were those three words I asked you to remember?"

I didn't know. Blank screen.

"Guess," he said. He smiled encouragingly.

"Apple, campfire, and...shovel."

"You have Korsakoffs."

I felt my eyebrows go up. All this time, all those visits, and this guy spits it out because I can't remember three words? "Korsakoffs."

"That's my initial diagnosis. I'm referring you to a neuropsychologist—not a shrink!"

"Not a shrink?"

He chuckled. "Everyone thinks a neuropsychologist is a shrink, but you're not crazy. A neuropsychologist can test things like memory problems, and is trained to recognize and diagnose disease without surgery or invasive tests."

"Korsakoffs. I had Wernickes."

"Wernickes can lead to Korsakoffs. It won't kill you, but there is no cure."

"I don't care if there's a cure or not," I said. I meant it. "I want to put a name to this thing so I can figure out how to deal with it."

"We'll set the appointment for you. The closest ones are in Ann Arbor, or Midland." Both are Michigan towns. Ann Arbor is a couple of hours easy drive away, and Midland fifty miles.

I left his office a nervous, but happy man. Finally, I had a tentative diagnosis that made sense.

§
Attempting to Prepare for the Neuropsychology Exam

The idea of letting someone probe my thoughts to see what was wrong with my brain scared me more than the tests that took pictures of my brain. I had fears. Fears my thoughts wouldn't be normal, fears my thoughts would have them lock me away in a rubber room. Little fears, and big fears. Ridiculous fears, and, I suppose, real ones.

Again I turned to the internet. I dug around to find out what kind of questions neuropsychologists ask, and what kinds of tests they perform. How did they know what was a normal ability to remember and what wasn't? I found very few details, and the general information I found wasn't very helpful. I saw one piece that said they build a few artificial tests into the battery to see if the patient is lying. They do that sometimes in criminal cases where the patient might very well want to lie about his memory, and I didn't think I had anything to fear in that way. I *wanted* him to find nothing wrong. I knew he would find something wrong, but I wanted him not to find anything wrong.

I found out the test was going to take all day. Just about fell out of my chair when I read that when I got the paperwork in the mail. *All day?* What the hell was that guy going to do to me that was going to take all day? Intelligence tests—IQ tests, and memory tests. All day. His office sent along a questionnaire for me to fill out and send in prior to the test.

I researched the guy who was giving the test. I'm not sure why I did, other than my quest for a list of questions he was going to ask was fruitless. I was able to find out the names of some of the tests and some general information about them.

That helped ease my fears. I was also able to find out where he lived, where he went to school, and when he got married. Scary what's available on the internet if one knows where to look. He seemed like a pretty good guy, and I hadn't met him yet.

All the literature I found said to get a good night's sleep before the neuropsyche exam. Easier said than done.

§
Returning to Midland

I love Midland, Michigan. I was the Boy Scout guy for Midland County from 1990 to 1997, and by the time I left, I served Midland, Gladwin, Clare, Isabella, and Gratiot Counties in either a direct or supervisory position with the Boy Scouts of America.

I got to town an hour before the exam and *by memory* and flawlessly, I drove by our old house, the homes of several volunteers, and the church we joined when we were first married. That's not as easy as it sounds considering we had lived in three different states since then, and I had learned my way around Chicago and Cleveland since then. Midland is not an easy city to drive around. There are a lot of one-way streets, and a lot of the side streets are only a few blocks long. I had no problem finding anything I wanted to find.

Memory problems! Ha! Maybe it wasn't Korsakoffs after all. Maybe I just hadn't been sleeping enough. That last bit rang true. When I was in housekeeping, I never slept more than four hours at a time. Twice a day I slept for about four hours at a time, but that's not the same as eight hours all at once. I got tired in the middle of the day, even after I left housekeeping. To this day (2011), I have a hell of a time staying awake all day, and when I do…I'm exhausted for three or four days after. Now I can sleep almost six at night, but I still have to have a couple of hours sleep during the day. It's related to Korsakoffs, I think.

The neuropsychologist was a skinny guy, and if he was over thirty years of age he hid it well. He struck me as arrogant right off the bat. I'm fine with arrogance, as long as it's earned. His seemed to be. I had no problem with the guy.

He opened by asking if I drove myself and seemed surprised when I said I had. I told him about my ability to drive to old haunts without problem. He said, "Classic Korsakoffs. Your memory prior to Wernickes is probably very good. What did you have for dinner last night?"

Blank screen. "No idea."

"You have Korsakoffs," he said. It sounded like a final judgment, and I was pretty sure he was supposed to test me before reaching a final judgment.

"Shouldn't you examine me first, and then diagnose me?"

"Your neurologist made the diagnosis. I'm just going to confirm it, and we'll see how bad it is."

"Fair enough." I was ready. I wanted to know. I *needed* to know. That crystallized the goal for me, and I'm comfortable with crystallized goals, even when I don't have a clue how to achieve them.

§
Intelligence Tests

Testing for depression is part of the neuropsyche exam. I don't remember much about that test other than it involved a bunch of questions, and that he was surprised to tell me I'm not at all depressed. "Remarkably well-adjusted," was what he said. I remember that much. That doesn't mean I didn't have times when I was recovering that I was down, and down on myself. I certainly did. I wasn't depressed in a clinical sense. I wasn't depressed before or after the neuropsyche exam. Quite the contrary—I felt I was finally in a place where I might get some information that would help me recover further.

I like IQ tests because you can't study for them. It's also darn hard to fail them. I was expecting a battery of written tests, like the ASVAB (Armed Services Vocational Aptitude Battery), although they claim it's not an intelligence test. I was expecting to have a highly qualified person, who, judging by the estimate of about $3,000 my health insurance provider said it would cost, watched me fill out a multiple choice exam with a #2 pencil.

It wasn't like that.

We sat in a small room that was more office than an exam room. We've seen enough television shows and movies to have an idea of what a psychiatrist's office looks like (but I don't know if they actually look like that or not.) This looked like an office. He sat behind a desk, and I sat in a chair in front of the desk. There were no windows in the room, and the closed door behind me didn't have a window either. There was a credenza behind the desk, and books on the credenza.

I had taken IQ tests in the past, and generally enjoy them. I'm not good at math. Never score well in math tests, and when it comes to mechanical things, I'm terrible. I knew those facts would come out in the tests. I also knew I would score well in any word test he wanted to throw at me.

...Post Wernickes, using the word "knew" takes on new meaning. I really didn't "know" I would do well on the verbal tests. I would have, before I was sick, but I wasn't all that sure I would retain the ability. I had read that Korsakoffs doesn't change IQ, but I had a little, nagging voice in my head—not as strong as Green Goblin's—that loved to plant, fertilize, and foster little, nagging doubts.

I find it highly ironic that I remember more about the memory tests of the afternoon than I do about the IQ tests in the morning. That might be because I did well on the IQ tests I thought I would do well on. He tossed the word "superior" around a couple of times on the word tests, and "very superior" once or twice. The results bear up to his reaction.

Conversely, the word "impaired" comes up when I look at the results for procedural IQ. The report states that I might use both halves of my brain for verbal-related tasks. There is an IQ disparity of 89 points between the two scores. The neuropsychologist called it "mildly lateralized." It means, in my case, that I use both hemispheres of my brain for verbal stuff. That might be a result of my childhood epilepsy, but we don't know for sure.

My memory of the memory tests might be stronger because parts of the testing frightened me. Wait. That's not quite strong enough. Parts of the memory tests scared the hell out of me. Yes. That's more like it.

§

I'm a Lefty—Take *that* Fourth Grade Teacher Who Tried to "Fix" Me!

We did a couple of tests to determine whether I'm left-handed or right-handed. One of the tests involved putting little pins in holes on a board. I had been playing around with doing things right-handed when I cleaned stuff for my employer. I learned quickly that I could clean the bathrooms faster if I used a Johnny mop in one hand to clean and a towel in the other hand to dry. I was trying to make myself ambidextrous, based on the faulty logic that as long as I was going to teach myself how to write, and even *think* again, I might as well try to figure out how to use my right hand for something other than typing.

When you're trying to get ready for work, and it's three AM when you're ready to shave your face and run out the door, it seems like a pretty good idea to hold an electric razor in each hand and shave your face. Never cut myself, not once. That's a joke, by the way. I've managed to do a lot of clutzy things in my life, but cutting my face with an electric razor wasn't one of them.

I flew through the test putting pins in the holes with my left hand. Then he had me switch hands. I would have done better if he didn't stop me every time I tried to slide the pin across the thing and let it hit a hole. He said it was cheating. I disagreed. I said that if the pin was in the hole and my right hand put it there, it shouldn't make a damn bit of difference whether the action was pretty or not. If that anal retentive neuropsychologist had been a referee in a basketball game, anything that touched the rim wouldn't have counted as a point. I didn't blame him, though. He didn't make the test.

He could see I was getting frustrated. Part of his evaluation dealt with how I handled emotions. I knew that, but that's not why I kept my cool. I didn't think I *was* keeping my cool. I was getting frustrated with him, the tests, the room, the pain in my legs, my situation…everything. There was a time or two that I sucked in my breath and let it out through my nose, only to proceed again without missing much of a beat.

On more than one occasion throughout the day he said, "God, you're well-adjusted." I wondered at the time how other patients handled their frustration. I thought I was a pretty emotional guy, but I'm not—at least according to every personality, depression, or other test anyone's ever tossed my way.

Know what else I'm not? Ambidextrous. It turns out I'm

left-handed. *Very* left-handed.

§
Lunch

I didn't really want to take a break for lunch, but I didn't want to deprive him of the break either. For me, the process was exciting. I felt I was finally getting close to getting to the bottom of what was wrong with me, and from the bottom I could start to climb up and out. I didn't think about the climbing I had already done. That's not my style, and maybe that was part of the problem. I'm always more focused on the mountain I'm on than the one behind me.

It was good to be back in Midland. I love Midland. There are a lot of good people everywhere, but it seemed like there were more good people in a small space than I had seen in other towns. Nothing against those other towns; it's just that Midland is that special. Besides… I wanted some time to drive by a few more of my old haunts.

I drove by the Little Chef restaurant, where I had lunch a couple of times a week. I didn't have time to go in, but wished I did. I was impressed by some of the changes, and equally impressed with some of the things that hadn't changed. I ended up eating at Taco Bell, not far from the neuropsychologist's office. I didn't want to be late getting back.

§
Memories, Tests, and Blank Screens
1
The List

Somewhere in this house there is a file. In the file is the document the neuropsychologist created with the results of my tests. I can't find it. That might not be bad. I was tempted to insert my scores on each of the tests he administered in this document, but I don't think that's what's important. I think what's important is for me to share what it felt like to take the tests and have holes in my memory come to light in ways I won't forget.

We did more than three memory tests, but I remember three of them because I failed miserably. "Fail" might be too strong, at least according to the good people who name the categories on the scores on the tests. I don't give their labels much credibility. They're the same people who thought "Very Superior" was a good combination of words to use to describe my verbal abilities. It's an annoying combination of words because it implies that there should be a category called "A Little Superior."

I can still hear his reedy voice as he read me a long list of words. The goal was for me to repeat the list to him after he finished reading it. As he read, I envisioned each word. I tried to find connections between them so I could paint a portrait in my mind with the intent of 'seeing' the portrait and 'reading' the words back to him from a script in my mind's eye. Know what I think? I think that when the people who make up these tests retire, they become golf course architects. They just can't shake the rat bastard out of themselves.

There was no way to paint a portrait from the random string of words he was tossing across the desk at me. It was like trying to read smoke signals on a windy day. They were clear across the sky for a few seconds, only to dissipate and blow away. One word followed another. I was getting frustrated and I hadn't even tried to repeat the list yet. I used to memorize plays when I was in college. I could nail everyone else's lines as well as my own. I never had much trouble remembering what I was supposed to say, and I never worried if someone else left the trail because I knew the plot well enough to help them come back.

Pppppplllllttt! Not anymore. Not then, at least. It's been a while on this road to recovery. I might be able to do theater again…but let's not get ahead of ourselves.

The time came all too quickly. "Repeat the list," he said.

I wanted a cigarette before I started. My nerves were jangled, and he told me at the beginning that if I wanted to go out to smoke, or needed a break, all I had to do was ask. He smiled and waved toward the door. I wanted the cigarette, but I was also stalling. I wanted to go outside to smoke. Smoke, and rehearse. Didn't know I was playing into his hands. I'm pretty sure he knew the time would work against me—I had time to forget words. Korsakoffs patients have short-term memory problems.

When I walked back in the room, I dove in without preamble: "Angel, forest, hill, sky, ribbon, butter…"

He held up a hand. "A little compulsive." He made a note.

"Bite me," I said with a smile. He didn't respond. I started again with the list, "Red, tree, table, box…" I ran out of words. It was frightening. One second they were there and then they were gone. Poof. The screen in my mind went blank.

"Take your time," he said in that flat drone of a voice he used. I wanted to tickle him to see if I could get inflection.

"It's gone." The muscles in my jaw clenched and unclenched. I looked at the books behind him. I looked *through* the books behind him. "They're all gone."

"Take your time."

"We can take a week, or a month. Gone is gone."

David J. Steele

2
Copy the Doodle

The drawing didn't look like anything. It was a line drawing of squares, rectangles, triangles, and shapes inside the shapes. It was, as the name of the test suggested (a name I don't recall in full), a complex drawing. He handed it across the desk and told me to copy it to a piece of blank paper. I could take all the time I needed to do it, but given the handedness test—I already knew it was a mistake to tell him I was trying to become ambidextrous—he suggested I use my *left* hand. I stuck my tongue out at him because I didn't know him well enough to give him the finger.

I drew it. Copied it. No big deal. He said I would copy it without looking at it later. I had no problem with that. I didn't trust my mental screen anymore when it came to pictures from lists of words, but I was cocky enough to trust it to remember a picture.

3
Story Time

He read me a short story, a piece of micro-fiction, maybe a page long. It was about a man who was going to leave his apartment to take a walk. After he read it, he asked me if I had any questions. I didn't. Then he asked if I wanted to take a break and go out for another cigarette. I knew what he was doing: giving me time to forget the story.

I went outside to smoke. It was getting hot. Summer day, blue sky. I smoked my cigarette and looked over a rusty fence and tall grasses at the traffic passing along U.S. 10. I forced my mind to relax and played the story over in my head like I was watching a movie. When I was done, I went back and repeated the story to him. I was pleased with my performance...until later.

4
Doodle—the Sequel

I thought he was going to ask me to repeat the story to him. He didn't. Instead, he slid a blank piece of paper across the desk and told me to draw the picture again.

I got a little frustrated when I tried. I got the overall shape right, immediately and without hesitation, and quickly—thinking I could do it before I forgot—drew the inner lines. Then I started to put in the random shapes. My mental screen went blank again before I was done. I knew there was more stuff, but it wasn't going to present itself to me. No way, no how. I stopped.

"Need more time?"

"Nope." I smiled, but I'm sure it was a sad smile. "That's all I've got."

"Okay."

5
Repeat the Story, Please

I can laugh about it now, but it wasn't funny then...my story was better than his story. I didn't embellish anything on purpose. That was against the rules. But embellish, I did. It's called confabulation—the brain filling in memory gaps with artificial memories.

In my version of the story, the man put on a raincoat. He put it on because it was cold, and drizzling. The weather report he was watching on television told him it was.

The problem?

The man *was* watching television, and he *did* go for a walk.

However…there was no mention of weather in the story *at all*. I had the story set in the evening, but there was no mention *at all* of time in the story he read me. I could have sworn he said there was. I could *hear* his voice when I played the story, and the scene, over in my mind.

Nope. He was right, of course, but I asked him to show me what he read. I trusted him, but I wanted to see the story with my own eyes.

For the first time in the process, I was frightened. I was scared. How could I have made things up without realizing it? More importantly, how could I trust myself not to do that in real life? Was I becoming a liar against my will? Was I going to disassociate myself from the real world a piece at a time, or wake up thinking I was Viper and actually, this time, *hurt* someone?

I asked for and received time for another cigarette break. Went outside and watched cars and trucks pass by on the highway. Tried not to cry or feel sorry for myself. It wasn't easy to avoid the temptation to hop in my car and go home without so much as a goodbye to the neuropsychologist.

I went back in.

6
Damn Doodle Again

He put another blank piece of paper on my side of the desk. I said…

"*YAAAAAAAAAAAAAAAH!*" and pretended to turn for the door.

Did he laugh, like a normal person would? Yes. Thank God, he did.

I picked up the pencil, sat, and started to draw. When the screen went blank in thirty or forty seconds, I was tempted to draw a big happy face with the tongue sticking out. Instead I tried to hide my shaking hands. The test was timed. A couple of minutes into the test I looked at him and grinned. "You're a schmuck. Anyone ever tell you that?"

"I've heard." He pointed at the paper.

"The screen is blank. That's all I have."

He made the paper go away.

He told me what I already knew about Korsakoffs—that there is no cure. He said my memory wouldn't get any better, but if I didn't drink it wouldn't get any worse. He suggested I find some sort of method of taking notes that would work for me and to make sure I was consistent with the method. I liked his suggestion that I pick up a digital voice recorder, and that's what I did. I still carry it with me.

One important note: I say I knew there is no cure for Korsakoffs, and it appears to be true. But that doesn't mean there can be no improvement in the condition. There *can* be improvement. I'm proof of it. It takes work on the part of the patient, and probably a lot of work on the parts of others. I don't need the voice recorder as much as I used to need it. I use it when I think I might need to have a memory. We'll talk more about the recorder.

For now, let's go back to the story...

§

Next Visit—this time with Sarah

It wasn't anyone's fault, and it cracks me up to think about it. It was a couple of weeks after my assessment, and my wife and I walked into the building.

Korsakoffs sufferers (and I was officially one) have memory problems. Ready for the laugh?

...They *remodeled the office!* Nothing was where it was

when I went there for my assessment. The decor was different. The hallway was different. The carpet was different, as were the pictures on the walls. As I held the door for Sarah, I looked at everything.

"...Went a long way just to mess with me," I muttered with a laugh. "Schmuck."

I was glad Sarah was with me. It's one thing to believe you have memory problems, it's one thing to know you have memory problems...but it's another thing completely to have a *medical diagnosis* that you have memory problems. I was resisting (and winning in my resistance) the urge to question everything I thought or did when it required memory. Every time I listened to my memories/reminders/messages on my recorder, I felt a tingle of what had become a real fear. I was afraid that one day I would push the button and have no memory of putting my memory on the recorder.

That never happened. I might not have remembered the reminder, but I always remembered *putting* it on the recorder. I made a rule for the recorder right off the bat: mine is the only voice on it. I don't let my wife make notes for me on the recorder. I don't record conversations or ask people to put reminders on my recorder.

It's not a digital voice recorder to me. It's a prosthetic. A prosthetic *memory* device. You wouldn't let someone stick a memory in your head, would you? I wouldn't. My recorder, my prosthetic memory, my voice.

We sat on a couch in the neuropsychologist's new office. I was glad the couch wasn't there on my first visit. I would have run screaming from the room. Her hand felt good in mine when he started to talk.

I was mostly quiet when he went over the results of each test. Squeezed her hand a little harder when the word "impaired" came into play. He gave us a thorough report with scores on each test and how I compared to the mean, etc. It's a useful and interesting report. It's also a little humorous to me that I can't find the damn thing now. It's around here somewhere. I'd love to blame Sarah's filing system, but I can't. Her filing system is actually pretty good. My guess is that report is somewhere on the bookshelf behind my desk, probably sandwiched between a cookbook I rarely use and last year's income tax return. I could look, but might cause an avalanche that would lead to my demise. No, thanks. I'll pass.

I didn't know how on edge I was until I started to relax. Perhaps I shouldn't have been nervous in the first place because I had a good idea of what the test results would say, and Sarah had been by my side all along. She's an educator, a good one, with more than fifteen years experience in special education. My IQ scores—the good, the bad, and the ugly—were no surprise to her. Neither were the memory deficits and struggles. It's hard to sit there and listen to people, even people who have your best interests at heart, talk about you in the third person as if you weren't sitting there. To hear your intellect and memory dissected and put into numbers is no fun. I wouldn't wish it on anyone. It's no fun, but it wasn't all bad either.

He gave me a copy of a rather lengthy report that covered all of the above, his recommendations including the digital voice recorder, and we paid the bill on the way out. I suppose I should qualify: we paid our share of the bill. Our medical insurance covered most of it. Our share was still a couple hundred bucks. It was worth it. With the diagnosis came peace of mind. With the prognosis—the idea that I wouldn't get any worse, but wouldn't get any better—came a challenge. I love a challenge like that. Wouldn't get any better? Bullshit.

David J. Steele

I felt pretty good when we left there. Sarah felt good, too. We had a name for the thing that had been slowing me down. Korsakoffs.

Section Five—Forward

§

Change in Sleep Needs

I get tired, exhausted, in the middle of the day. It's a heavy exhaustion. I usually go with it and sleep for at least an hour and a half, but prefer to crash for two, three, and sometimes four hours. That's when I tend to dream and remember the dreams. At night, I can't seem to sleep for longer than six hours. It's usually less than that. I have found no evidence that confirms my belief that my change in sleep habits is a result of Korsakoffs. That doesn't change my belief. If I don't or can't —for whatever the reason—get my afternoon sleep, I pay for it for about a week. I feel drained, almost ill.

I don't know if anyone is researching sleep habits of Korsakoffs patients. Frankly, there are bigger fish to fry when it comes to Korsakoffs. If we need to sleep, we can and should sleep. I think the bigger fish to fry should be fried (to carry an analogy to the point of uselessness) trying to figure out how to get Korsakoffs patients back on their feet so they can get back into the world.

We *can* get back into the world. I'm not the only one who's done it. I might be a noisy one, but I'm not the only one.

§

Voice Recorders Make People Nervous

My employer has a policy that specifically prohibits recordings of any kind in the workplace. My voice recorder clearly was in violation of that policy. I wasn't going to give it up, but felt the need to do the right thing and let my boss know I had it and that it was prescribed to me by a medical professional. I'd been using it for three weeks, but now that I had a diagnosis and the orders in writing, it was time to mention it. I left my supervisor a voice mail message the next night I worked.

The next day I worked, I was called in to her office. The voice recorder was okay if I was able to provide in writing from a doctor that I had to have it. I wasn't insulted by that, and acquiesced with no problem at all. When they pushed further and asked that I leave the voice recorder behind when I left work so they could listen to it at random and make sure I wasn't stealing company information or eavesdropping on others, I got mad. I didn't say anything, but my posture—and the smoke coming out of my ears—as I left was pretty easy to read.

I went home and got out the report. We still have the fax machine I bought when I worked at home a lot, and still have a photocopier that works. I made a copy of the report from the neuropsychologist and blacked out the IQ results. I had good reason to black out the IQ results—I was working in the accounting department, and although I didn't need to add or do any actual accounting other than data entry, I thought it wouldn't be good to provide conclusive proof that I'm impaired in mathematics. (Chuckles. They knew it already, but knowing and proving are two different things.)

I faxed the pertinent portions of the report, specifically the part that said I should use a camera, voice recorder, or notepad. Further, I rebelled against the suggestion that I leave my voice recorder behind for two very good reasons: 1) it's a *prosthetic*, for memory as opposed to something like a mechanical arm, but a prosthetic nonetheless; 2) It was *my* recorder. I didn't ask the company to pay for it.

After the fax went through, I went online and found a copy of the American Disabilities Act and read it thoroughly. Sure enough, my voice recorder was covered. It was a needed device to assist me in doing my job. When I put the printed copy in a folder with my test results, I hoped I wouldn't have to use it. Lawsuits aren't my thing, but neither is getting punished for following medical advice. I left the folder at home when I went to work the next time.

The Human Resources Manager made a point to talk to me. She apologized for the whole conversation about the device, and did so nicely and before I said anything. I was allowed, of course, to keep and use my little recorder. She hadn't thought of it as a prosthetic before and apologized for the idea that I would be asked to leave it behind. I offered to let her listen to it at any time if they felt the need to make sure I wasn't recording anyone else. I also pointed out that the device doesn't pick up anything more than a few feet away unless I equip it with a microphone. You have to hold it close to your mouth to record your own voice.

No one has given me any grief about it since then. Few know I have it. I don't like to use it when anyone else is around.

§
When You Don't Know What You'll Forget…

…how do you know what to record?

Trial and error, mostly. Initially, I made a note whenever something seemed important and I didn't want to forget it. I used to have trouble remembering what hours I worked so I could calculate how much my next check was going to be. The position I was in had specific start times, but every shift ended at an unspecified time—whenever the last customers left and the bartender finished counting their cash was when I left. I used the recorder to make a notes to myself with the hours I worked. I used the recorder to make notes to myself with To Do stuff, and To Done stuff. I used the recorder a lot. I would love to tell you that I found specific patterns to things I forgot. I didn't. The forgetfulness is random. Annoyingly random.

It's a double-whammy, that forgetfulness. It's not a matter of not remembering. It's a matter of *not knowing what you do or do not remember.* Can you understand the difference? I hope so. When I forget something, it's as if the memory never existed in the first place.

It's been three years since I started carrying a voice recorder. They're not very expensive and can be found at the local RiteAid or Walgreens. They're digital and you don't lose messages if you forget to change the batteries.

I lost one of them. That was funny, more than it was frightening. I had a spare. More than one person said, "Where did you last have it?" Helpful question, right? ...Not so much. My reply when asked where I last had the recorder was: "If I knew where I last had it, I wouldn't need it." It's a good line. It holds enough truth to make a point, and it usually gets a laugh.

Yeah. Sometimes life is like that.

§

Improvement. Is My Brain Healing?

Web sites, conversations with doctors, reading medical articles, and participating in discussion groups about Wernickes seems to show that most healing from Wernicke-Korsakoff (they call it *improvement* instead of *healing*, but I'm not going to split those hairs), occurs in the first six months to three years.

I would break my improvement/healing down to three time periods: a) in the hospital and recuperative care, b) in the six months to three years immediately after, which is about the time between my release and the neuropsychological assessment, and c) *after* three years.

My memory has improved quite a bit. I don't blank things out the way I did before. I still have slips, mostly when I'm tired—and my memory goes out the window, relatively speaking, when I'm sick. For the most part, though, it's pretty close to normal...with a few additional blank spots. Sarah and I were joking about one of those blank spots last week.

Most of the time, she picks up toilet paper. I told her it's because I hate to buy the stuff. I thought it was true until Sunday afternoon when I had no choice but to go to the grocery store to buy toilet paper. I'm not picky about the stuff, but she has one clear preference. I try to buy the one she likes; I really do. She doesn't ask much out of life, but when she wants something I want her to have it.

I never, I mean *never,* buy the right kind of toilet paper. Seriously! I write down the kind I'm looking for. I've put the brand in my little recorder. Every time, when facing that long aisle of toilet paper—with the variations within the brands, and the brands, and the store brands, and the colors... Blank screen. I doubt my note, doubt my recorded message...and buy the Charmin.

Charmin isn't the right kind, but somehow I convince myself it is. The hell of it is, even *I* don't prefer that stuff, and I don't care that much about toilet paper brands. I do think it's a Korsakoffs thing.

I might be off the hook now. I finally 'fessed up and told

her about my blank. My toilet paper blank.

On the one hand, not knowing what kind of toilet paper to buy isn't a big deal. Raise your hand if you're a husband and can't remember what brand of toilet paper is on the roll in your bathroom. Uh-huh. That's what I thought. Don't sweat it. Screw it up enough and you won't have to buy it anymore. So I'm forgiven the toilet paper thing.

Anyway… I think this Sunday's toilet paper confession is a good sign. It's a good sign in that I have finally identified something I can't remember *that's consistent.* All my other memory lapses have been pretty random. Now that I know and have admitted to myself that I can't remember what brand of toilet paper to buy, I can start looking for other consistencies in my forgetfulness. Maybe then I'll be able to identify a pattern and come up with coping mechanisms that work. And, no, I'm not going to have "Charmin" tattooed on my body.

§
Brain Work

I read in several places (though, ironically but not surprisingly, I can't remember where) that there is a theory a person with brain damage will start using other parts of their brain to accomplish things. It might be true, it might not be true. I'll tell you this: *The more I write, the better my memory seems to be.* I suspect it's because of the way my brain is set up, but the point is that mental exercise helps. What doesn't help is riding the river and giving up.

A couple of years have passed since I sat in my mother-in-law's house looking through the windows at the river below and typed books others had written. I've self-published seven books since then. This one will be number eight. The more I write, the more I use all parts of my brain, and the more I do that, the better I remember.

I've been asked how I can write novels with multiple plot lines when I have memory problems. Here's how: I write one chapter at a time. When I finish writing a chapter, I print it, then retype it. If it's the end of the day when I re-type the chapter I just wrote, I type it again in the morning. By the time I've gone over the chapter the second time, it's locked. When I first started doing that, I needed to retype the chapter three or four times, sometimes, to lock it in. Now twice is enough. I might be able to get away with writing one or two chapters in a row before retyping, but I still do it. Either way—whether I still need to do that or not—it's a good habit to revise and correct as I go.

§ Sobriety

I remember what the doctor I mentioned earlier in the book said about having one or two being okay. I'm not going to do it. In my case, it isn't okay. I've "slipped" once or twice, but not often. The slips are followed by a period of self-anger I don't like. I don't do that anymore. On the couple of occasions when I tested myself and had a beer, I paid for it with bouts of dread and uneasiness in 24-48 hours. It's not worth it.

I've had foods prepared with alcohol in the recipe: a beer dumped on hamburgers prior to grilling, in a rum cake, and maybe one or two recipes others prepared. Same deal as above: uneasiness, and random twinges of unidentified fear. Now I ask if there is alcohol in the preparation of food, and if there is, I have something else.

That doesn't mean there aren't times when I would love to sit down and drink a few beers. There are times when it sounds great. Then I don't do it. I just don't. I have too much to lose, and I'm not going back to that dark place with the red hills. Sobriety is key. I still have no desire to attend AA meetings, though I know they help a lot of people. I do my best when I don't think about drinking, and do other things with my time, like write, and cook, and bake, and read, and spend time with my wife, and, and, and…

I still enjoy situations where alcohol is present. I like bar environments, and restaurants, and parties. I actually like them better now that I'm sober. Some people who have had struggles with alcohol and alcoholism don't or can't be around those situations, and most people respect that. I think I'm lucky I can go anywhere and not drink. If it bothered me I would avoid those situations, and that's probably not easy. In my present, paying job (other than writing books), I spend hours at a time less than ten feet away from the bar. Not only does it *not* bother me, I like it.

You see, my friend, every time I don't drink…every day I'm sober, every night I fall asleep sober…I win. When I go to the grocery store or drug store, I go out of my way to walk past the beer, wine, and booze. If you're next to me you might hear me hum and chant, "Nee-ner-nee-ner-NEE-NER!" I'm singing to the containers. I won.

I'm not going to play again, mind you, but I won.

www.ingramcontent.com/pod-product-compliance
Lightning Source LLC
Chambersburg PA
CBHW021955170526
45157CB00003B/998